Taking Stock

THE CENTENARY HISTORY OF
THE ROYAL WELSH AGRICULTURAL SOCIETY

Taking Stock

THE CENTENARY HISTORY OF
THE ROYAL WELSH AGRICULTURAL SOCIETY

David W. Howell

Published on behalf of the Royal Welsh Agricultural Society

UNIVERSITY OF WALES PRESS
CARDIFF
2003

www.wales.ac.uk/press

BRITISH LIBRARY CATALOGUING-IN-PUBLICATION DATA
A catalogue record for this book is available from the British Library.

ISBN 0–7083–1825–8

The Royal Welsh Agricultural Society wishes to acknowledge the financial support of HSBC in publishing this volume

Designed and typeset by Andrew Lindesay at the Golden Cockerel Press Ltd.
Printed in Great Britain by J. W. Arrowsmith, Ltd.

Contents

The Royal Welsh Agricultural Society has much to celebrate in its centenary year. After overcoming many early difficulties and setbacks in its history, over the last two decades the society grew into a vigorous and successful organization. Its annual show as the shop window for Welsh farming has become recognized as one of the three major agricultural events in the United Kingdom, and its out-of-show activities embrace a wide-ranging programme to promote the agricultural and rural economy of Wales.

This book to celebrate the centenary, written by Professor David Howell who is an acknowledged expert in the history of rural Wales, gives readers a comprehensive picture of the evolution of the Society, tracing its development from its first six years at Aberystwyth between 1904 and 1909, through its nomadic phase from 1910 to 1962, to its permanent home at Llanelwedd from 1963. With the help of illuminating and sometimes entertaining anecdotes, the centenary history presented here is above all a human story; a record of the wonderful voluntary support given by countless people in different capacities to ensure that the Society carried out its founders' aims of promoting Welsh agriculture and, above all, its world-famous livestock. I am sure that readers will finish the book filled with a sense of pride and gratitude for the faithful service rendered to Welsh farming communities since 1904 by the Royal Welsh Agricultural Society.

As a visitor on a number of occasions to the showground at Llanelwedd, which is set in one of the most beautiful parts of the country, I have been keenly interested in the fortunes of the Society and the farmers that it is there to serve. In particular, I have been very aware of the vital role the Society has played in supporting Welsh farmers through some of agriculture's darkest days in the last decade, most notably Foot and Mouth in 2001. I will never forget the atmosphere at the Royal Welsh Winter Fair in December 2001, the first gathering of farmers after Foot and Mouth – the sense of relief was tangible, as was the determination of the remarkable farmers of Wales to continue in the way of life to which they had been born and in which so many excel. The Royal Welsh Agricultural Society has much to be proud of and I could not be more delighted or honoured to write this foreword for 'Taking Stock'. I hope and pray that the Society's next 100 years will see it develop to meet the new challenges that lie ahead, and in so doing, grow from strength to strength.

Author's Preface and Acknowledgements

For someone who had throughout his academic career written on the development of Welsh agriculture and the nature of Welsh rural society from the eighteenth century onwards, I was thrilled (though not a little nervous) to be invited by the Royal Welsh Agricultural Society in 1997 to write its centenary history. My first visit to the Society's show was as a schoolboy when, in July 1955, Narberth Grammar School ran a trip to the highly successful Haverfordwest show, the year in which the event for the first time was covered by television. Little did I realize then that the Society's annual shows, marvellous occasions as they were for the parading of many proud lords of the ring, were far from being the sole *raison d'être* of the Society's existence. I hope that readers will discover through these pages its many-sided activities, all in the pursuit of promoting the well-being of Welsh farmers and of the wider rural community. At various times following its stormy inception at Aberystwyth in early 1904, and particularly in the years before the mid-1970s, organizers were confronted with many difficulties that threatened the institution's very existence, and an awareness of the struggle to surmount them successfully should make Welsh people all the more proud of the achievements and international reputation of this great Society.

In the course of writing the book I have received valuable help and guidance from many institutions and individuals and I take pleasure here in acknowledging my indebtedness. As always, the staff of the National Library of Wales, Aberystwyth, eased my path; in particular, Menna Phillips of the Department of Printed Books, herself the proud daughter of Llywelyn Phillips who played such a key role in the Society's affairs in the 1960s and 1970s, placed her considerable knowledge at my disposal. The permission given by the head of the Manuscripts Department at the National Library, Gwyn Jenkins, for those minutes of the Society deposited there to be temporarily transferred to the library of my college in Swansea was a great help to me. The staff of the Thomas Parry Library at Aberystwyth were always courteous in allowing me ready access to their almost complete run of the Society's *Journal*. My colleagues in the library at the University of Wales Swansea once again showed their customary willingness to assist in every way possible. Very patient indeed were the staff of one of Wales's most elegant buildings, the round reference room of the city library, Swansea, in helping me to use microfilm copies of many back numbers of the *Western Mail*. W. Dyfrig Davies of Teledu Telesgop, Llandeilo, provided suitable photographic material for the cover of the book. My many inquiries about details relating to a wide range of the Society's activities, past and present, were always answered promptly and courteously by the permanent staff at Llanelwedd. Chief executive David Walters and his team there have made me feel at home during my periodic visits to the showground. The information provided by the Board of Management, under its chairman, Dr W. Emrys Evans, together with its permission to borrow minutes of the Society and other relevant material hugely facilitated the preparation of this volume.

In all my writings on Welsh rural history stretching back to the 1970s I have benefited from the advice of two leading scholars in British agricultural history, Professors Gordon Mingay and Michael Thompson. Conversations with them about the approach to take in writing this centenary history were of great value. My reliance on the scholarship of Professor Richard Moore-Colyer of the Department of Agriculture at the University of Wales, Aberystwyth, has been extensive. Of course, the annual shows have always featured Welsh cobs and ponies and my knowledge and understanding of this important branch of the Society's activities were vitally underpinned by both the writings and correspondence of Dr Wynne Davies, the acknowledged expert in this field. In addition, my friend and research student, Wilma Thomas of Llanmadoc, Gower, a past competitor at Llanelwedd with working hunter ponies, also helpfully commented on early drafts relating to the equine section. Some readers will remember the lively personality of Alan Turnbull of Gower who played a big part in organizing the shows at Llanelwedd. I gained insight into his character through meeting as a group a number of his Gower friends and I owe thanks to Judy Methuen-Campbell of Penrice Castle for arranging this meeting at her house. This book has benefited enormously from the skills of my friend Alun Owen of Llangyfelach, who spent many hours scanning a large number of old, impaired photographs dating back to the early decades of the twentieth century. I owe him special thanks. Valuable help was given, too, in scanning other old photographs by my colleague Roger Davies of the Arts and Humanities photographic section at University of Wales Swansea. Equally vital was the assistance given me in translating Welsh texts into English by Brinley Jones, Ifor Rowlands, Dr Peter Freeman and Dr Gareth Pritchard, colleagues of mine. Another colleague, Nick Woodward, son-in-law of the late Wil Jones, a past president of the Welsh Pony and Cob Society, kindly constructed figures 1 and 2 in Appendix Three. The completed text was read by Professor Geraint H. Jenkins of the Centre for Advanced Welsh and Celtic Studies at Aberystwyth and by Dr Jeremy Burchardt of the Rural History Centre, University of Reading. Both made valuable suggestions towards improving the finished product.

My college has given me magnificent encouragement. Vice-Chancellor Robin Williams, FRS, a native of Bala, has always enthusiastically enquired after the progress of the book. While my head of department, Professor Noel Thompson, and other colleagues in the History Department have given over-whelming support and shown exemplary patience towards my preoccupation with the Royal Welsh project, the really hard work fell to our secretary, Jane Buse, who typed the book with her customary cheerfulness and efficiency. I am greatly indebted to her. It is a pleasure to record the Society's and my thanks to the University of Wales Press, especially Susan Jenkins, Ceinwen Jones, Ruth Dennis-Jones, Liz Powell and Sue Charles, for their unfailing courtesy and guidance at every stage in the publishing of both the Welsh-language and English-language volumes. My thanks go also to Janet Davies who compiled the indexes in both languages, and to M. Eluned Rowlands for the translation of my original into Welsh.

Finally, I have received constant support from my wife, Angela, and our daughter, Emma Angharad, during the researching and writing of this book. As always, I have cause to be very grateful. My late father, too, took a huge interest in the book's progress and I shall always remember the delight he showed at being told by me at his Kilgetty home that the book had been completed.

David W. Howell

Picture Acknowledgements

The author and publishers acknowledge the following sources and individuals for permission to reproduce illustrations:

RWAS archives, including illustrations from the *Royal Welsh Agricultural Society Journal*: 1, 2, 3, 9, 10, 11, 12, 15, 16, 17, 18, 19, 20, 21, 22, 23, 24, 26, 27, 32, 34, 36, 37, 38, 39, 41, 42, 43, 44, 47 (Les Mayall) 49 (Tegwyn Roberts), 50, 51 (Tegwyn Roberts), 52 (Tegwyn Roberts), 53 (Tegwyn Roberts), 54 (Tegwyn Roberts), 55, 56, 57 (Tegwyn Roberts), 59 (Tegwyn Roberts), 60, 62 (Tegwyn Roberts), 63 (Tegwyn Roberts), 64 (Tegwyn Roberts), 66, 67, 68 (Tegwyn Roberts), 69, 70 (Tegwyn Roberts), 72 (Tegwyn Roberts), 73 (Tegwyn Roberts), 74 (Tegwyn Roberts), 75, 76 (Tegwyn Roberts), 77 (Tegwyn Roberts), 78 (Tegwyn Roberts), 82, 83 (Tegwyn Roberts), 84 (Tegwyn Roberts), 85, 86 (Tegwyn Roberts), 87, 88 (Tegwyn Roberts), 89, 90 (Tegwyn Roberts), 91 (Tegwyn Roberts), 93

The Thomas Parry Library, Aberystwyth for the following items from the *Welsh Journal of Agriculture*: 4, 5, 6, 7, 28, 46

Griffiths & Davies, Dolclettwr Hall, Taliesin: 8

Central News, London: 13

Shirley and John Thomas, Sketty, Swansea: 14

National Library of Wales, Geoff Charles Collection: 25, 29, 30, 35, 40, 79

Megan Thomas, Pennard, Gower: 31

Hammonds, Hereford: 61

Dr Wynne Davies, Miskin, Pontyclun: 33, 65

Miss Biddy Gwynne Howell, Llanelwedd: 45

Western Mail and Echo: 58

Marina Gallery, Llandrindod Wells: 48, 71

Daily Express: 80

Farmers Weekly: 81

John Kendall: 92

A Message from our Sponsors, HSBC

HSBC Bank plc is delighted to be able to sponsor this book which celebrates the centenary of the Royal Welsh Agricultural Society.

Most people will associate the Society with the annual Royal Welsh Show which is held in July and indeed it was the first show in 1904 that laid the foundations for the current Society. The influence of the Society is now much more far-reaching with a number of specialist shows and events held throughout the year on the Society's permanent show ground at Llanelwedd.

Remarkable progress has been made since the show ground was acquired in 1962 and the daily activity is a testament to the hard work of many willing individuals from across the whole of Wales. The Society's work attracts many people to mid-Wales during the course of the year and is a major contributor to the local economy.

Agriculture remains a significant part of the Welsh economy and the Society provides a focal point for many sectors. The shows in particular encourage excellence in breeding and stockmanship which are essential for Welsh produce to compete in world markets.

HSBC's association with the Society goes back for many years and we are delighted to be able to add our own congratulations on achieving 100 years of progress. We look forward to working with the Society for many years to come.

PART ONE

The Society is Launched

THE ABERYSTWYTH YEARS, 1904–1909

A Long Wait

The foundation of the Breconshire Agricultural Society in 1755 by a local squire, Charles Powel of Castell Madoc, created the first county society of its kind in Britain, which drew its inspiration from the London Society of Arts set up in the previous year for the encouragement of agriculture, manufactures and commerce. In an attempt to improve methods of husbandry practised within the county, premiums were awarded to progressive farmers for growing turnips and potatoes, free turnip seeds were provided, trees were planted and roads were widened and improved. Influenced by its beneficial effects on the farming of Breconshire, the aristocracy and gentry elsewhere in Wales established similar societies in their own counties. First to follow the Breconshire example was Carmarthenshire when, in April 1772, Watkin Lewes of Abernant-bychan (Cardiganshire) – who also owned estates in Carmarthenshire – proposed the institution of a 'Society for the Encouragement of Agriculture, Planting, and other laudable purposes'. Later in the same year a society was also founded in Glamorgan under the patronage of Thomas Mansel Talbot of Margam. The Pembrokeshire Society for the Encouragement of Agriculture, Manufacture and Industry was set up twelve years later by William Knox, the newly arrived squire of Slebech, who had earlier in his career been Under-Secretary of State for America. For reasons unknown, the society lasted a mere six or seven years, but it was revived in 1805 by Lord Cawdor of Stackpole Court. That same year, 1784, also saw the founding of the Society for the Encouragement of Agriculture and Industry in Cardiganshire. Six years later there occurred the metamorphosis of the London Radnorshire Society, which had met at the Gray's Inn Road Coffee House, into the Radnorshire Agricultural Society under the encouragement of John Lewis of Harpton Court. Both the Wrexham and the Montgomeryshire Agricultural Societies were founded in 1796, the former by the third Sir Watkin Williams Wynn of Wynnstay, and societies were later set up in Merioneth, Caernarfonshire and Anglesey in 1801, 1807 and 1808 respectively.

Even if the impact of these early agricultural societies fell short of the aims of their founders, they certainly did assist in the spread of new methods of husbandry such as growing turnips and clover. In the early years membership was restricted mainly to landowners whose periodic meetings held at local inns were characteristically convivial and bibulous, but from around 1810 it was increasingly extended to include substantial tenant farmers. Of significance in this respect, the Pembrokeshire Society, in its drive to recruit a larger membership from among ordinary farmers, had a standing order from 1809 that 'dinners shall not exceed half a crown each person, and that everyone present will be permitted to chuse [sic] his own beverage'. While no longer, therefore, the sole participants in these societies, the aristocracy and gentry continued their support in the decades that followed. If that patronage was not forthcoming societies collapsed, as happened, for example, in the case of the North Cardiganshire Agricultural Society in 1885. A fresh impetus was given to existing societies in the 1860s when landowners and the more progressive farmers saw the value in improving the breeds of Black Cattle and other livestock at a time when prices were moving upwards. Many new societies were also founded at this time, a phenomenon which was observed by *The Welshman* in 1874: 'The rapidity with which these exhibitions are multiplying in the country is a convincing proof of their popularity, and no doubt the fame of one society, or the knowledge of the good accomplished in one district, has led to the formation of other societies.'

Other contemporary acknowledgement of the beneficial influence of these societies is plentiful. In 1882, for example, local landowner and cleric, the Revd Garnons Williams of Abercamlais, attributed the improvement in agriculture in Breconshire over the past years to agricultural shows. Clearly the pronouncement of the squire of Cilgwyn, Cardiganshire, in 1872 that small local agricultural societies 'really do no good and that it is a waste of money to encourage them' was too sweeping. Nevertheless, certain qualifications have to be made about their effectiveness. Tenants, it was claimed, valued agricultural shows mainly for the prize money which they could win and failed to realize that the true object of a society was to give landlords and tenants not money but good stock. On occasions they grumbled if the landlords won the best prizes, and to meet this objection prize lists were drawn up with primary consideration given to the best means of financially rewarding the largest number of exhibitors rather than the best way of improving stock and crops. A novel solution to the problem was attempted by the Merioneth Agricultural Society in 1881 when it decided to divide the showyard into two parts, one for stock of tenants with incomes below £300 a year and the other for landowners' stock; whereas money prizes were awarded to tenants, landowners received merely honorary rewards. That this was a successful experiment is suggested by the adoption of the practice by the North Cardiganshire Agricultural Society the following year. The swift growth in numbers of these societies in the late century was also criticized, with one commentator claiming in 1884 that such highly localized organizations were weak and that it would be better to have a small number which were strong in both resources and influence.

Despite all this early activity in founding local agricultural societies from the middle of the eighteenth century onwards, Wales was the last of the United Kingdom countries to form a national agricultural society. The first, the Dublin Society, was set up in 1731 by Thomas Prior to encourage agriculture, manufactures and the arts, and it received a good deal of support from the great Irish landlords. This was followed by the Highland and Agricultural Society of Scotland in 1783, which obtained a Royal Charter in 1787. There is no mistaking the fact that it was a real source of inspiration for the founders of the Royal Agricultural Society of England – principally John Charles, the third Earl Spencer, and William Shaw, the editor of the *Mark Lane Express* – who brought that society into being in 1838. There soon followed the founding of the North-East Agricultural Association of Ireland in 1854, whose formation was spearheaded by the fourth marquis of Downshire known as 'The Big Marquis'. This society had some connection, however tenuous, with the North East Society, set up in 1826. In 1903 the North-East Agricultural Association changed its name to the Ulster Agricultural Society, and the following year it was granted permission to style itself the Royal Ulster Agricultural Society.

As the opening chapter will reveal, the founding of the Welsh National Agricultural Society would encounter the jealousies of existing local agricultural societies and long-standing animosities between the inhabitants of south and north Wales. Similar difficulties had meant that Black Cattle breeders of the south and north went their own ways in the mid-1880s in forming separate herd books and institutions, a state of affairs which would remain until the eventful year of 1904 when the two societies amalgamated to form the Welsh Black Cattle Society. It is significant that both the Welsh National Agricultural Society and the Welsh Black Cattle Society were launched at a time which was astir with developments geared towards improving Welsh farming. We can cite the continued growth since their establishment in the late 1880s – not least in their extension work in the form of dairy schools – of the two Departments of Agriculture at the University Colleges of Wales at Aberystwyth and Bangor; the founding of the Welsh Pony and Cob Society in 1901; and from 1902 the growth of the cooperative movement in south-west Wales pioneered by Augustus Brigstocke. Indicative of the shared personnel of the various bodies, two active supporters of the cooperative movement, D. D. Williams and Walter Williams, were also prominent in the founding of the Welsh National Agricultural Society.

CHAPTER ONE

Petty Jealousies Overthrown

The continuing success of the Royal Welsh Agricultural Society as it celebrates its centenary year in 2004 could hardly have been anticipated by the handful of individuals who founded the Welsh National Agricultural Society a hundred years earlier. Indeed, during its first year or so the Society faced downright hostility from a range of determined opponents, and for many years afterwards the organizers would be hindered and frustrated by a chronic lack of members which threatened the financial stability of the Society to the point of its very survival. The heroic efforts and stout, visionary commitment of those early founders cannot be appreciated without a detailed examination of how the Society was set up and how it overcame the considerable opposition to it.

1. Sir Lewes Loveden Pryse, the first director and secretary of the Society from 1904.

The Welsh National Agricultural Society was the brainchild of Lewes T. Loveden Pryse of Aberllolwyn, Llanychaearn, near Aberystwyth, the extravagent, debt-ridden and supplicant third son of the great landed house of Gogerddan who would fortuitously inherit the family acres and baronetcy in 1918. In January 1904 he was in his thirties and was secretary of the Llanfarian and District Agricultural Co-operative Society. It is possible that inspiration had come from the resounding success of the North Cardiganshire Agricultural Society's show at Aberystwyth the previous year at which he, along with W. B. Powell of the nearby Nanteos estate, and Vaughan Davies of Tan-y-bwlch, Aberystwyth, MP for Cardiganshire, had guaranteed £360 for the open prizes. Although unpopular with the farmers of the district, these open classes attracted cattle to Aberystwyth from all counties in Wales except Radnorshire, and even King Edward VII had sent stock to demonstrate his interest in Welsh agriculture. It was soon apparent that gentlemen exhibitors from across Wales were keen to build on the success of the 1903 show, and so Loveden Pryse resolved to elevate the event at Aberystwyth to the status of a national show for Wales. This strategy first came to public light at the annual general meeting of the North Cardiganshire Agricultural Society on 1 February 1904 at the Lion Royal Hotel, Aberystwyth, when he proposed that the local classes should henceforth be held at the existing show at Tal-y-bont and at a newly created one south of the

River Rheidol at Llanilar. This course of action would clear the way for a large open show at Aberystwyth. Loveden Pryse had already approached certain people whom he considered to be representative of agriculture in Wales, and the meeting was informed that the following had all registered their support for the initiative: the earl of Powis, who agreed to accept the office of president; Lord Tredegar, who consented to becoming one of the vice-presidents; J. Marshall Dugdale of Llwyn, Llanfyllin, R. M. Greaves of Wern, Porthmadog, and Richard Stratton of The Dyffryn, Newport (the three Welsh representatives on the Council of the Royal Agricultural Society of England), although the last-named while welcoming the new venture declined to join the Society on account of his not being a Welshman; A. Osmond Williams, MP for Merioneth; A. C. Humphreys Owen, MP for Montgomeryshire; Colonel Pryce Jones, MP for Montgomery Boroughs; and Vaughan Davies, the aforementioned MP for Cardiganshire. We shall see that, after heated discussion, the proposal that a big open show should henceforth be held at Aberystwyth was carried.

Encouraged by this response, Loveden Pryse then called the following gentlemen together at the Lion Royal Hotel, Aberystwyth, on 11 February 1904: Daniel Davies Williams (who took the chair), Captain G. Checkland

2. D. D. Williams, Argoed Hall, Tregaron, a prominent founder member of the Society.

Williams, Henry Roberts, J. R. Rees of the local branch of the North and South Wales Bank and Rufus Williams. Daniel Davies Williams, or 'DD' as he became familiarly known, the son of a wealthy farmer of Tregaron, was on the staff of the Department of Agriculture at the University College of Wales, Aberystwyth and director of the College Farm, Tan-y-graig. There is a good case for arguing that 'DD' should be accorded equal status with Loveden Pryse as joint-founder, so enthusiastic was he for the success of the new venture and so important were his considerable business acumen and love of detail in the preparation of the Society's original aims. Fluent in Welsh and English, he gave his all to the Society for a round half century until his death in 1954, a fortnight before the Society's jubilee show at Machynlleth. Upon forming themselves into the Welsh National Agricultural Society and formally electing the earl of Powis as president and Lord Tredegar as one of the vice-presidents, they proceeded to draw up some draft rules to be submitted for approval to the first general meeting of the supporters. As the House of Commons was sitting and many MPs wished to be present, this meeting, attended by more than twenty prominent figures from Wales, was held in London, in Committee Room Twelve at the House of Commons, on 26 February 1904.

It was proposed that the following rules be passed. First, that the Society should be called 'The Welsh National Agricultural Society', with Monmouthshire being included as part of Wales, and that its aims should be to improve the breeding of stock and to encourage agriculture throughout Wales. With this end in view (expressed by the Society's organizers in both Welsh and English), the following schemes were to be undertaken just as soon as the Society was in a position to do so:

1) To hold an annual show whose object was to get the best of stock from all parts of the United Kingdom exhibited in Wales so as to give the Welsh farmer an opportunity once a year of seeing the best types of British breeds and, by giving special prizes confined to Welsh farmers, to encourage them to breed their distinctive breed up to the best standard.

2) To start new, and assist existing, Entire Horse, Pony and Bull Clubs throughout Wales, an aim which, the founders were mindful of, followed the lines of the Royal Ulster Agricultural Society.

3) To edit an agricultural journal in Welsh and English for distribution, free of charge, to members of the Society.

4) To assist agriculture generally throughout Wales in every way in its power.

The second rule proposed and for which approval was sought was that all questions of a political tendency relating to measures pending or to be brought forward in either House of Parliament should be absolutely excluded from discussion at any of the Society's meetings. There then followed a number of proposed rules bearing on the Society's constitution. First, that it consist of a president, vice-president, trustees, governors, and members; second, that the trustees and vice-presidents could only be elected from the class of governors; third, that the Council consist of the president, not more than thirteen trustees, thirteen vice-presidents, and fifty-two members; fourth, that the president be an annual officer who should not be eligible for re-election for three years and that the election of the president and Council should take place at the annual general meeting and that they should enter into office at the conclusion of that meeting; and, finally, that subscriptions be set at £5 for governors and £1 for members – with the exception of bona fide tenant farmers or occupiers who made their living by farming in Wales and whose rateable value did not exceed £100 per annum, whose subscription should be 10s. 6d.

These were all carried, upon which it was proposed 'that the Show be held at Aberystwyth for the next three years at the expiration of which time the question of locality to be decided at a general meeting of the members' and it was also resolved 'that the Show be open to the world with a few special prizes confined to bona fide tenant farmers or occupiers, making their living by farming in Wales'. Several proposals followed (all of which were carried) for the election of certain people, mainly large landowners and some MPs, to the Society's various offices and bodies such as the Council, Finance Committee and the Schedule and Judges Selection Committee. Predictably, Lewes T. Loveden Pryse was elected general manager and he immediately proposed that the show be held as a one-day event on 3 August 1904.

A major consideration behind the proposal to hold an annual national show was the recognition that whereas Welsh breeds had, in most cases, received limited recognition in showyards outside Wales, they had, owing to the multiplicity of competing breeds, been unable to obtain that especial and particular recognition an essentially Welsh national show would be able to offer them. Moreover, such annual shows would afford an opportunity to the prize-winning animals from local shows to compete, thereby ensuring a representation of the best in Wales, which in turn would draw together both good producers of stock and intending purchasers, the results of which would beyond doubt be beneficial to Welsh agriculture as a whole.

In favour of the establishment of a Welsh National Agricultural Society with its annual show was the appreciation in the minds of the founders that Wales had always provided a satisfactory venue for visiting agricultural shows. Thus when the Royal Agricultural Society of England had held its show at Cardiff in 1901 it proved, they contended, to be one of its very few meetings of late years to be an unqualified success. (Certainly they were correct in so far as the attendance of 167,423 was concerned.) Or, again, when the Bath and West of England Society had visited Swansea in May 1904 the same happy outcome attended it, they claimed. The founders were sanguine that such success could be easily repeated by the Welsh National Agricultural Society, thus bringing considerable benefits to Welsh farmers. In their minds was the example of Scotland with its hugely successful Highland and Agricultural Society.

Initially, there had been some misapprehension that the show would be permanently held at Aberystwyth. The founders were thus keen to stress that this was only a beginning and that the combined circumstances of a generous donation and the offer of a good site on which to start such an enterprise had led to the agreement that it would be advisable to hold the show in the same place for the first three years. However, at the end of that period the question of where to hold the show would be referred to the decision of a general meeting, or placed in the hands of a Council representative of the various districts and breeding societies. The founders were sustained by the conviction that if the idea were to take hold of the Welsh people and win widespread support from north to south, the Society would soon be a prosperous undertaking.

This general support was, in part, forthcoming from the outset. Many well-known landowners and agriculturalists in Wales, as well as twenty-seven of the thirty-three Welsh MPs, supported the Society from an early stage.

Walter Williams of Aberystwyth, assistant secretary of the Society, was to write in the *Western Mail* for 21 May 1904 how 'the landed gentry have responded splendidly'. Yet there was powerful opposition from parties jealous of the initiative shown by Lewes Loveden Pryse and his immediate circle. R. M. Greaves of Wern had exactly predicted this sour response when he wrote to Loveden Pryse in February 1904:

I am very glad to hear that there is a prospect of a National Welsh Show, which is a thing that is greatly wanted, and which I have long hoped to see established, though I had always feared that the petty jealousy of the local Societies would be an insurmountable obstacle.

Bitter indeed were the sustained attacks lasting down to 1906 on Lewes Loveden Pryse by John Gibson, the editor of the *Cambrian News,* whose offices were at Aberystwyth. His quarrel was that Loveden Pryse's high-handed action in setting up the Welsh National Agricultural Society at Aberystwyth would inevitably lead to the extinction of the existing successful North Cardiganshire Agricultural Society and its annual show which had always been held at Aberystwyth. Loveden Pryse had vexed Gibson at the annual general meeting of the North Cardiganshire Agricultural Society at the Lion Royal Hotel, Aberystwyth, on Monday, 1 February 1904 by proposing that since the present show had become such an important event it would be advisable in future to hold the local classes at a new show to be set up south of the River Rheidol and at the present show at Tal-y-bont, north of the Rheidol, thereby enabling Aberystwyth to have a big open show. He noted that there was already £1,000 in hand, and that, as we have mentioned, some prominent Welsh agriculturalists and MPs from certain mid-Wales constituencies had registered their support.

In his response, Gibson urged the farmers present to speak like men on their own behalf and say what they really thought of what he denigrated as the very 'curious proposal to form two sorts of shows, one in the south and one in the north; while somebody of whom they knew nothing held a show in the middle of the district with which they had nothing to do'. He moved as an amendment that consideration of the proposal should be postponed for a fortnight to enable farmers and other subscribers to the North Cardiganshire Agricultural Society to have time to think about the matter. It seemed to him, he defiantly submitted, that the proposal was intended to kill the North Cardiganshire Agricultural Society in order that one person should be allowed to 'boss the county'. Not surprisingly, there followed an ill-tempered exchange between Gibson and Loveden Pryse over the question of whether or not Gibson was a member of the North Cardiganshire Agricultural Society, which Gibson ended by saying: 'I am not going to pay you for bossing this Society.' As the *Cambrian News* for 5 February tersely reported: 'At this stage of the proceedings there was considerable excitement and a general conversation was carried on.' Tempers were further raised when W. B. Powell, squire of Nanteos, intervened to say that Gibson's remarks were most insulting to Pryse, a comment which drew approval from the meeting. Continuing indistinctly, ran the same report, Powell said that after having brought the North Cardiganshire Agricultural Society into its present state it was unfair for Gibson to refer to Pryse in such a blackguardly way and he ought to be kicked out of the room. At this point the chairman called for order and Gibson's amendment was seconded by D. Rees of Tynpark. Loveden Pryse, for his part, refuted criticism that the people of the district would not have a voice in the management of the show by stressing that it would be run on the same lines as the Royal Agricultural Society. There was no question of his bossing, for there would be many bosses over him. The Society had a council composed of twelve vice-presidents and fifty members, and tenant farmers would have an opportunity of electing the fifty members. No one could boss a society of that kind, he added to applause.

In response to a question from John Jones, Ynishir, as to whether the local farmers would be eligible to compete in the Aberystwyth show, Pryse replied that anybody would be able to compete in the Welsh Agricultural Show and that there would be several special prizes offered to Welsh farmers alone. Pressed as to whether the North Cardiganshire Agricultural Society would hold its own show at Aberystwyth that year, Pryse responded that, although the Welsh Agricultural Show would be held at Aberystwyth instead of the usual event staged by the North Cardiganshire Agricultural Society, anybody would be eligible to compete for the prizes.

After prolonged lively discussion the question was put to the vote. Twenty voted for Loveden Pryse's motion, twelve for Gibson's amendment. Despite his defeat, the editor of the *Cambrian News* was in no mood to accept the situation and successive issues of the newspaper were to see him employ his formidable rhetorical skills as an editor to castigate the new venture and its founders. He immediately turned on Osmond Williams, MP for Merioneth, and the other members of Parliament who had pledged their support, arguing that they would not have countenanced such a high-handed piece of action in their own counties as had happened in Cardiganshire where a new show was being established by a few individuals connected with the North Cardiganshire Agricultural Society and at the cost of its destruction. How would Osmond Williams, he asked, as a keen supporter of the Merioneth Agricultural Society, feel if one of the officials of that society produced a cut-and-dried scheme for the burial of the Merioneth Society and the apportionment of some of the funds between two side exhibitions, one at

Harlech and one at Corwen, in order that a show open to the whole of the United Kingdom should be held annually at Dolgellau? He then sneered: 'We think that the official who made a proposal of this kind in Merioneth would be very peremptorily asked if he had not forgotten who he was and what his duties were.' In berating those MPs who had promised their support to this 'mad-hatter' scheme, he questioned whether they knew how it had originated, how it did not enjoy the support of the agriculturalists of north Cardiganshire, and how it necessitated the destruction of the existing North Cardiganshire Agricultural Society.

A week later Gibson returned to the fray in issuing the dark warning that the 'irresponsible autocrats' who had set up the scheme had made the radical mistake of proceeding irregularly and had established side shows at Llanilar and Tal-y-bont without regard to the rules. Indeed, they had generally acted as if the subscribers to the North Cardiganshire Agricultural Society were imbeciles and of no account. For Gibson, perhaps the most extraordinary fact in connection with the wrecking of the North Cardiganshire Agricultural Society by the 'autocrats' was that they did not even appear to have taken the trouble to consult the Council of the United Counties Society, which included Cardiganshire, Carmarthenshire and Pembrokeshire. He went on to assure the Council of that society that the North Cardiganshire farmers had nothing to do with the 'idiotic' proposal which was the collective imagination of a very small number of 'megalomaniacs'. This tirade of 19 February 1904 ended in a hysterical, sour rant:

During the last 40 years we have seen a good many shadowy Welsh national movements of individual origin. They have all come to nothing. This show for the whole universe will also come to nothing. What the farmers and landowners of North Cardiganshire have to do is to see that the agricultural society [North Cardiganshire] which, until last year was successful, does not also come to nothing.

All his declamations boiled down to one basic argument, namely, that no national society, however successful, could compensate the farmers of North Cardiganshire for the loss of their local society.

Gibson was not the sole critic of the new venture. At Carmarthen on 10 February 1904 a meeting of the United Counties Agricultural Society (which had been founded in 1896) passed a resolution condemning the organizers for starting the so-called Welsh National Agricultural Society without prior consultation with the existing older societies in Wales. On 2 April a further meeting of the Society at the Boar's Head Hotel, Carmarthen, chaired by its president, C. Morgan-Richardson of Neuadd Wilym (Cards.), voiced renewed criticism of the founders of the new society. The president informed those present that after the resolution of protest had been passed at the last meeting the secretary, D. H. Thomas of Starling Park, had forwarded a copy to several societies and that he had received notice that some of them had already adopted a similar kind of resolution, amongst them the Pembrokeshire Society and the South Wales Society of Black Cattle Breeders. Responding to claims that the action taken by the United Counties Society sprang from jealousy, Morgan-Richardson countered that its objection to the scheme stemmed from the improper way it was set up. He stated bluntly that the Aberystwyth show could never become a national one, and that it could only do some degree of harm to the United Counties show whilst doing no real good to itself. No leading landowner in south Wales, he claimed (though somewhat inaccurately given the support of Lord Tredegar, Colonel Wyndham-Quin and Sir Henry Fletcher), had joined the Society and, in his view, it was noticeable that no representative from Carmarthenshire or Pembrokeshire had attended the first general meeting in the Committee Room of the House of Commons. The fact was, he urged, that the show at Aberystwyth would be of no value whatsoever to any farmer in those two counties.

Morgan-Richardson denied that his strong criticism of the project at the February meeting had been motivated by personal or private reasons, as claimed by Loveden Pryse in a letter to their secretary. In that correspondence Pryse's main objection to the United Counties Society's resolution of protest was because it accused the North Cardiganshire Agricultural Society of forming the Welsh Agricultural Society whereas, he insisted, the new society had been formed by influential men throughout the whole of Wales. So upset had he been about a circular issued by the United Counties Society's secretary on 18 February making this same accusation against the North Cardiganshire Agricultural Society that he had issued a warning that sending out these circulars rendered the United Counties Society liable to legal proceedings, and he called upon the secretary to contradict this statement wherever he had circulated it. Morgan-Richardson declared to the meeting that there was nothing to fear from this threatened libel action, and, revealingly, he went on to observe that it was now a case of Carmarthen against Aberystwyth as an agricultural centre. Continuing in defiant mood and fed by his audience's approval, he predicted that if Carmarthen town were to get improved rail facilities and if their members stood true to their own society then the future need not be feared; indeed, what they had imagined would harm them would rather redound to their advantage as it would bestir them to make their show even better than it had been in the past.

After some discussion it was proposed that a copy of the resolution passed at the meeting of 10 February should be sent to all existing Welsh agricultural societies and their opinions on it canvassed. In short, this resolution expressed the United Counties Agricultural Society's view that the inaccessible position of Aberystwyth made it unsuitable as the place for a central show and that the committee of the North Cardiganshire Agricultural Society had no right to form a national society to represent the whole of Wales without consultation with the older agricultural societies already in existence. Despite some support for the Aberystwyth venture from Major Webley-Parry-Pryse of Noyadd Trefawr and David Evans of Llwyncadfor stud farm, the motion was carried unanimously.

Gibson's detestation of the new venture latched onto this 'reasonable action' taken by the United Counties Agricultural Society in bringing the whole matter before every agricultural society in Wales. Similarly, the *Western Mail*'s editorial for 12 April 1904 came out in support of the United Counties Society's attitude, baring its conservative fangs in observing: 'The very name by which it is proposed to call the Society shocks everybody's sense of propriety. The term "national" has been grossly abused of late years owing to the loose way which it has been employed.' Further, the intention of holding the proposed Aberystwyth 'national' show for one day only was ridiculed as unlikely to attract exhibitors from a distance. On the other hand, a letter carried in the *Western Mail* of the same date and written by a 'supporter' lambasted the objection of the United Counties Society, putting it down to that 'petty jealousy which has been the curse of so many good causes in Wales'. It was all about, he concluded, Aberystwyth stealing a march on Carmarthen as an agricultural centre, a matter simply of local jealousy.

In response to the United Counties Agricultural Society's request a meeting of the Lampeter Agricultural Society, for example, was held at the end of April 1904 chaired by J. C. Harford of Falcondale. Early in the proceedings a letter from Lewes Loveden Pryse, dated 28 April, was read out by the chairman in which the writer claimed that the secretaries of many agricultural societies had written to him for particulars of the new national society and that many of them had rendered him great assistance. 'How very different', he observed, 'their conduct is from that of the United Counties who without asking for any particulars, burst forth through their President into an exhibition of childish opposition and endeavour to kill what they themselves allow to be a good movement, because they were not consulted.' Given that the main object of the new movement was to encourage livestock breeding by the farmers of Wales, Pryse wrote that he was confident that the Lampeter Society would not oppose the venture, and he stressed the view that, far from hurting local shows, a large national show would improve them. Differing views were put by those present as to whether the new movement should be supported. The United Counties' claim in its circular that it was the action 'of certain gentlemen of Aberystwyth' was criticized by several speakers. However, the chairman supported the United Counties' complaint that there had not been consultation with existing societies and went on to add that Aberystwyth was a wretched place to which to send stock. Two strong supporters of the venture at the meeting were David Davies of Velindre and D. J. Williams of Aber-coed, the former urging those present that they should not quarrel about the name which the new society had assumed and that recognition should be given to the greater number of classes at the big show to be started at Aberystwyth, an event 'which would benefit all'. When the proposition that the Lampeter Agricultural Society should approve the formation of a national show to be held at Aberystwyth was put to the vote, it was carried unanimously.

On 29 April 1904 another meeting was held to discuss the movement, this time at Pwllheli, appropriately with Loveden Pryse and Walter Williams in attendance. It appears that none of the objections that had been forthcoming from the south were voiced by the big societies in the northern counties and the meeting at Pwllheli – if but a small gathering with less than thirty in attendance – certainly registered no dissonance. Its purpose was to explain the objects of the Welsh National Agricultural Society, such a society according to John Greaves, the lord lieutenant of Caernarfonshire, having long been 'the despair and ardent desire' of well-wishers of Welsh agriculture. Not the least of the obstacles facing the Aberystwyth initiative, Greaves correctly observed, would be to induce north and south Wales and even the counties in those divisions to pull together. He was followed by Loveden Pryse who, in pressing the objectives of the Society, made the point that success at local shows often led to false conclusions as to the quality of breed and a big show of the kind instituted would remove these false notions and encourage them to better things. After Pryse had sat down, Walter Williams of Aberystwyth, the assistant secretary, spoke in Welsh and referred to the National Agricultural Societies of Ireland and Denmark in which, he said, national pride was taken. To his listeners' approval, he asked why Wales should not have a similar society and a similar pride. R. M. Greaves of Wern spoke to observe that as a member of the committee of the Royal Show (England) he was made repeatedly to deplore that there was no agricultural authority in Wales to which the committee could refer; such an authority would be invaluable in ventilating the grievances and articulating the opinions of Welsh agriculturalists. In an attempt to address the widely felt concern that the new society would injure the smaller Welsh ones, R. M. Greaves also made the point that a national society would improve rather than harm them. Furthermore, he believed that such a society would be of huge benefit in bringing together north and south Wales. The debate was then joined by J. Bryn Roberts, MP for north Caernarfonshire, who remarked that the smallness of this gathering at Pwllheli should not discourage the promoters, for the beginning of nearly all successful institutions, as the movement for the Welsh University Colleges forty years ago, was similarly wanting in public interest, and he concluded his address with an appeal to farmers and agriculturalists to join the new society. The meeting ended by unanimously approving the formation of the Welsh National Agricultural Society.

The debate continued in agricultural circles in the south. Notably, there emerged in early May 1904 a resolution by the United Counties Agricultural Society to bring all bodies together to thrash out the question of forming a truly National Agricultural Society for Wales. Meeting at Carmarthen on 3 May the United Counties Society resolved that its secretary, D. H. Thomas, and the secretary of the Glamorgan Chamber of Agriculture, D. T. Alexander of Cardiff, be asked to convene a joint meeting of representatives of every agricultural society in Wales during the Bath and West of England Show to be held in Swansea later in the month. The meeting would discuss the question of a National Agricultural Society for Wales and Lewes Loveden Pryse, the general manager of the projected 'national show' at Aberystwyth, would be invited to explain the aims and objects of the newly formed Society.

The *Western Mail* editorial for 20 May welcomed the holding of the meeting at Swansea on the morrow. After gently chiding the Aberystwyth initiative with the observation 'It may be that the promoters did not set about the task in the way best calculated to ensure success at first', it observed in positive vein that 'much may be accomplished by a friendly discussion of ways and means. If the movement which has been started at Aberystwyth is to go on – and there is every reason to believe it will – the sooner all agricultural bodies in Wales and the new Society come to an agreement the better. It is only by perfect understanding and co-operation the new venture can be made truly "national".' Noting that one of the questions to be discussed at Swansea was whether the new show was to be stationary or migratory, the *Western Mail* came down in favour of a migratory one, citing the Court of the University of Wales and the National Eisteddfod Association as demonstrating the virtue of an itinerant institution.

On 21 May, the room at the Bay View Hotel was packed with representatives of agricultural societies and others in south Wales. The meeting, chaired by Colonel Lewis of Llysnewydd, Carmarthenshire, heard Morgan-Richardson once again criticize the founders of the Aberystwyth movement ('two or three people at Aberystwyth') for failing to consult the existing societies of north and south Wales; he complained that the founding meeting was not held in a public place but rather it was a private one in the House of Commons and that, furthermore, the rules were objectionable. If it was to be a national society, Morgan-Richardson contended, the exhibits should be confined to the people living in the Principality, thereby protecting the farmer against whom he chose to call the 'travelling showman'. Again, Aberystwyth was a most inaccessible place, and it was not even in a good agricultural district: 'They could have nothing to exhibit there – except mountain ponies, probably.' A more positive note was nevertheless struck in his announcing that they (presumably meaning the agriculturalists of the south) were prepared to support a show which would move from place to place. The resolution he then went on to propose embodied this approach:

That in the opinion of this meeting the show recently established at Aberystwyth as a national show for Wales has been formed without consultation with the existing agricultural societies in North and South Wales, and that only a movable show to be held from time to time in convenient centres in North, Mid and South Wales will be generally acceptable to the agriculturalists of the Principality.

Loveden Pryse responded in commendably conciliatory mood, stating that in his opinion the motion at the London meeting for the show to be held at Aberystwyth for three years could be rescinded in favour of just one year, and he denied that the scheme had been pushed by Aberystwyth people. Indeed, the earl of Powis had been deliberately chosen as president because he resided in central Wales, and the proposals were supported by people from all over Wales. If it was a mistake not to consult the agricultural societies it was one for which he alone was solely responsible and he expressed himself 'very sorry' if any one had been offended. That he had acted so was to save time, for to have consulted everyone would have meant inevitable delays. Equally conciliatory in its turn, the meeting refused to pass a resolution with any reference to the action of the Aberystwyth founders and it resolved that all that needed stating was that a national show would be acceptable only if it was migratory and that a small committee be nominated to consult with Loveden Pryse as to the future development of the movement. A motion was also carried that two representatives of each agricultural society of north and south Wales be asked to meet Loveden Pryse at a central place to discuss the scheme further.

Gibson's continuing ill-temper burst forth into print in the *Cambrian News* on 27 May. While he conceded that a Welsh Agricultural Society might ultimately be established, 'the one-man movement' against which he had protested from the outset was 'as dead as the proverbial doornail'. For it was made clear at the meeting, he sneered, that the 'Welsh Universal Show' would probably not be held at Aberystwyth more than once; that the price of next year's show there would be the practical destruction of the North Cardiganshire Agricultural Society; and that if a Welsh National Agricultural Society were to be established it would not be managed by one man and would not be always held in the same place. He took obvious pleasure in reminding his readers that in these days 'it is not as easy as it used to be to speak and act without authority in the name of the people of Wales' – a clear swipe at the old aristocratic regime. Nor is there any mistaking his satisfaction in pronouncing that the 'Swansea meeting on Saturday pricked the bladder of the Welsh National Agricultural Society and utter collapse followed'. One response to Gibson's thundering came from Colonel G. F. Scott of Plas Cregennan, Arthog, Merioneth, who wrote to Loveden Pryse at the end of May:

I am greatly amused at old Gibson's leading article. It reminds me of the Russian despatches, after a Japanese victory. I consider, all things considered, you won all along the line at Swansea. It is only one more instance of local and personal jealousy. You ought to be most warmly thanked.

In agreement was the response of the *Western Mail*, although in judging the Swansea meeting a success it was at pains to commend the wisdom of the proposal to make the show migratory and surmised that the promoters of the newly formed society would agree with this. It pulled no punches in declaring that a stationary show at Aberystwyth could never thrive, for most Welsh agriculturalists would simply fail to support a show held from year to year in a remote corner of Wales. As far as the editor was concerned, Wales was too divided to co-operate in any movement that was more or less local in its scope, once again citing in his defence the fact that every national organization in Wales was migratory like the Cambrian Archaeological Association, the Court of the University of Wales and the National Eisteddfod.

By the time that the first show was held on 3 and 4 August – the original proposal for a one-day show had been revised, perhaps in response to criticism that this compared unfavourably with the many two- and even three-day shows of the English counties – much of the earlier antagonism towards the National Agricultural Society had evaporated. The *Western Mail* editorial for 3 August remarked approvingly, 'Since the date of the concordat which was arrived at between the agriculturalists of the joint counties at Carmarthen matters seem to have gone on very smoothly' (a reference to an understanding reached at that town on 6 July which had agreed to confirm the resolution passed at Swansea). Such a disarming of all serious opposition was a tribute to what the newspaper chose to describe as the 'persistency and thoroughness' with which the original idea of Lewes Loveden Pryse had been pursued. A non-editorial in the same newspaper on the following day commended his having 'laboured manfully and won the day'. To squeals of protest from the older societies for not being consulted in the formation of the new organization, pointed out the writer, was offered the reply that 'too many cooks spoil the broth' and that initial agreement would have been almost impossible if every existing society had been asked to take part from inception. Rather, it was hoped to win general support gradually by virtue of the proposal's own inherent comprehensiveness. According to this correspondent, if at the end of three years there was a widespread feeling that the interests of the National Show would be enhanced by changing the venue and making the exhibition a migratory one, an arrangement had been reached by which that wish would be realized.

The *Western Mail* was by now seeking to reconcile the United Counties' members to the National Show, using the success of the United Counties' own August show at Carmarthen to argue that there was no reason why both shows should not exist and flourish side by side. But to no avail, for opposition continued from both Gibson and the United Counties Agricultural Society. The *Cambrian News* for 23 February 1906 thus saw Gibson smirk that this 'pretentious' society was not paying its way, which did not surprise him because people in those days were not going to subscribe money freely to a 'one-man society' with a number of titled and other sort of figureheads who had no real influence. Once again he returned to his hobby-horse: 'It is an artificial product and will fizzle out in the end unless it sinks back into its original form as a North Cardiganshire Society under popular control.' In spiteful exultation he continued: 'We have waited three years for the present humiliating position. We can wait twice three more years for the final collapse.' Writing in the *Cambrian News* of 27 February 1906, Sir Richard D. Green-Price was moved to rebuke Gibson thus: 'I have failed to discover why a local paper such as yours should have gone out of its way to cast a stone at this Society, which is without doubts conferring benefits not only on Aberystwyth, but also on the whole of Wales.'

As part of its continuing opposition, late August 1904 saw the United Counties Agricultural Society suggest that Glamorgan should be invited to amalgamate with it, the successful outcome of which would firmly establish the United Counties as the south Wales show. Those supporting this idea argued that the show at Aberystwyth met the needs of north Wales only, and that it could develop itself in that direction, whilst leaving the Carmarthenshire amalgamated show to spread its influence over the south Wales area. According to United Counties' president, Morgan-Richardson, the idea was not new, merely the revival of a similar proposal made some years earlier. Although he claimed that this move was not intended to come into opposition with 'the Aberystwyth or North Wales Show', it is difficult to accept this at face value. Again the *Western Mail* for 22 November 1904 adopted a conciliatory role in urging the executive of the proposed amalgamated society to see to it that as few stumbling blocks as possible were placed in the path of the Welsh National Agricultural Society: 'That organisation, now that it has been established, should be given a chance to live and prosper.' In the event, the subscribers to the old Glamorgan Agricultural Society decided by April 1905 not to amalgamate with the United Counties Agricultural Society and this seems to have been the end of the matter.

By the time the second show was held in early August 1905, there was a sense that the initial prejudice against the institution had failed to inflict lasting damage. Indeed it was argued in the *Western Mail* that such opposition from local agricultural organizations had only served to advertise the national show, that 'the wind which blows against the Aberystwyth kite only serves to lift it up'. Just one last attempt to throw a spanner in the works was to come at a meeting of the Council of the United Counties Society in December 1906 when the prize-list committee recommended that the Society be extended to include south Wales and Monmouthshire. If this were to be achieved, it was urged, a liberal prize list would be the means of promoting greater improvement of livestock, of materially adding to the popularity of the Society, and of increasing the entries at the show. Favourably disposed to the motion, the meeting resolved that the following year the show at Carmarthen would be open to the counties of Monmouth, Glamorgan, Radnor and Brecon as well as to Carmarthen, Cardigan and Pembroke. After three years of this extended open show, however, it was felt that since west Wales provided most of the prize money it was better to confine the show to the three south-western counties. The new venture thus withstood the jealous tantrums of local societies in the Aberystwyth district itself and in the south. Moreover, on Gibson's part there was unmistakably a personal animosity nursed towards Loveden Pryse, fuelled, it appears, by resentment towards the old aristocratic families.

CHAPTER TWO

The Show Gets on the Road

There is no doubt that, despite the opposition to the new venture, the first six years of operation laid a firm foundation for the Welsh National Agricultural Society's show. Indeed, by the end of its period at Aberystwyth D. D. Williams, writing in the *Journal of the Welsh National Agricultural Society* for January 1910, could justifiably reflect: 'The Welsh National, although only six years old, has already reached the forefront of agricultural shows.' He recognized that, although this mere fledgling of a society had not yet reached the state of perfection attained by other long-standing national societies, a vast improvement had been noticeable annually. Nevertheless, this indefatigable worker on behalf of the Society was well aware that certain defects had to be removed. Thus the stewarding of the show in many sections was 'feeble', and for a national show more use, he felt, should be made of the experience of secretaries of Britain's leading shows and other well-known men. Award boards were not used in any of the rings, and all too often visitors simply could not follow the judging. Furthermore, certain alterations were desirable in many of the classes: a class should be provided for hackney foals and, in so far that in the 'Sheep other than Welsh' section many Ryelands were exhibited, an effort should be made to establish special classes for this largely Welsh breed. On balance, however, strengths far outweighed weaknesses, and the

Table 1. Livestock entries, 1904–1909

SECTIONS AND CLASSES	1904	1905	1906	1907	1908	1909
Shire horses	45	55	46	65	44	51
Hackneys	31	36	29	34	28	44
Hunters	19	19	23	22	27	25
Ponies and cobs	111	96	101	104	121	118
Shorthorn cattle	68	68	63	70	55	60
Hereford cattle	25	47	46	60	56	50
Welsh Black cattle	56	74	64	65	52	40
Shropshire sheep	29	41	41	46	23	52
Kerry Hill sheep	36	44	56	82	63	59
Welsh Mountain sheep	46	64	56	73	64	70
Pigs	16	16	16	47	15	9
TOTAL ENTRIES	482	560	541	668	548	578

organizers were greatly encouraged by the fact that the number of livestock entries showed an overall upward tendency over these first six years, as can be seen in Table 1.

Of immense and understandable satisfaction, also, was that these figures exceeded the entries in the Bath and West of England Show. From the very beginning, one of the distinctive features of the Welsh National Agricultural Show was the substantial proportion of total exhibits falling within the tenant-farmer classes, almost 50 per cent in 1904 and just over 60 per cent in 1905, with these, encouragingly, coming from all parts of Wales. In an attempt to secure further tenant-farmer involvement in the shows, the Society decided to give second prizes in all of the special classes open to tenant farmers. The editor of the Society's *Journal* had, in fact, recommended this in 1905, enthusing that:

Having prizes confined to Welsh farmers is an excellent institution, as it gives them a chance of competing with the professional exhibitor, and comparing their animals with the best the country produces, and gives them encouragement by being able to compete amongst themselves for a prize.

Further grounds for satisfaction came from the fact that exhibitors sent stock from all quarters of England as well as from Wales, their number including several winners at the Royal Agricultural Society of England Show. Of particular importance in arousing public interest were the exhibits of Hereford cattle sent year after year by King Edward VII although such was the high standard of entries that, as in the 1907 and 1909 shows, they did not always take first prize.

As was to be expected and certainly intended, to the fore in the livestock sections at the successive Aberystwyth shows were the specialities for which Wales was famous, namely, Welsh Mountain ponies and cobs, Welsh Black cattle, Welsh Mountain sheep and Kerry Hill sheep. Most popular of all, and the greatest crowd pleaser, were the ponies and cobs. Outstanding in the Mountain pony class entries was Grey Light, a stallion belonging to prominent Welsh pony breeder Evan Jones of Manoravon, Llandeilo. Victorious in 1904, 1905, 1907 and 1909, Grey Light was considered to be amongst the finest ponies in the kingdom. From the 1908 show onwards the competition received a further stimulus in the form of the annual presentation by the prince of Wales (who had become a patron of the Society the previous year) of a £50-silver challenge cup for the best cob of the old Welsh type from four to seven years old, not under fourteen hands or over fifteen, bred and owned by a bona fide resident of Wales and Monmouthshire, with the competitors to be eligible for entry or entered in the Welsh Pony and Cob Society's Stud Book. It was wholly appropriate that the Welsh National Agricultural Society selected as judge in this key competition of the show Sir Richard D. Green-

Price of Presteigne, an experienced judge and breeder and an enthusiastic advocate of the pure Welsh cob, whose outspoken desire to keep the breed

3. Grey Light, the celebrated pony stallion, four times winner at Aberystwyth, belonging to Evan Jones of Manoravon, Llandeilo.

undefiled by the hackney strain was well known. Since the object of the Prince of Wales Cup was to promote the breed of the old Welsh cob, Sir Richard decided at the 1908 show to select out of the sixteen entries a mare best capable of producing the stock desired, with the prize accordingly going to H. P. Edwards of Aberystwyth's Pride of the Hills. In 1909 fourteen animals were entered for the competition, of which some ten were rejected on the grounds that they were too much of the hackney type. The winner, High Stepping Gambler 2nd belonged to Evan Davies, Cwmgwenin, Lampeter, and was the only real specimen of the old Welsh cob in the whole show. It was hoped that awarding the prize to a stallion would give a special impetus to the breeding of Welsh cobs now so much in demand at home and abroad.

With their established reputation as a milk- and beef-producing breed, and in its capacity as the rent-paying stock of many tenant farmers in Wales, it is hardly surprising that Welsh Black cattle were a popular feature of the show from the beginning. Indeed, the very existence of a big show for Wales was paying off in so far as the exhibiting of Welsh Blacks was concerned, with the judge at the 1905 show observing that one point was beyond all question, namely, that the show of Black Cattle at Aberystwyth was infinitely stronger in both numbers and quality than at Park Royal, near London, the temporary permanent home of the English Royal Show between 1903 and 1905, after which disastrous episode peripatetic shows were resumed. One commentator was to enthuse that Aberystwyth's 1906 show featured the best lot of exhibits of Blacks seen anywhere. Competition between north and south Wales was very keen, as in 1908 when, having already won his class at both the previous shows at Aberystwyth, the celebrated bull of cooperating farmers Messrs

Davies, Thomas and Howells of St Clears, Duke of Connaught, carried away the Ysguborwen Challenge Cup for the best bull of the Welsh breed and a number of other special awards, thus beating competitors from north Wales, although the latter got the better of the south in the female classes. The annual show was also a means of tenant farmers' obtaining really good prices for their

4. Duke of Connaught, the Black bull which won at Aberystwyth between 1906 and 1908.

stock, including Black Cattle. At the 1905 show, for instance, some Welsh Black calves fetched £20 apiece. Thus the big show in Wales, as elsewhere, acted very much like a large fair where the best buyers congregated.

By virtue of its annual shows the Welsh National Agricultural Society not only provided a stimulus to the improvement of the breed, but also worked in harmony with the newly formed Welsh Black Cattle Society, which had come into being in early 1904 through an admittedly reluctant marriage – reflecting, so J. B. Owen of Llanboidy, Whitland, maintained in the *Journal* for January 1905, 'Southern prejudice against the Northern breed' – of the formerly separate North Wales and South Wales Black Cattle Societies. Conveniently, the annual general meetings of the Welsh Black Cattle Society were held at Aberystwyth during the time of the show, as were those of the Welsh Pony and Cob Society founded in 1901, and the Welsh Sheep Flock Book Society.

The founding of the latter during the 1905 show, with Marshall Dugdale, a true lover of the real Welsh breed of sheep, at its head, was urgently needed, and further encouragement to the improvement of the breed was provided by the Welsh National Agricultural Society's annual show. There were claims that the exhibition of Welsh Mountain sheep at the 1905 show was the finest that had ever been seen anywhere, and it was this display which prompted the immediate calling of the meeting at the showground to establish the Welsh Sheep Flock Book. There was much work for all native-breed

25

enthusiasts to do, as was demonstrated when the judge of the Welsh Mountain sheep in the 1907 and 1908 shows drew attention to the fact that alongside excellent entries of Welsh Mountain sheep there were still animals submitted that were not worthy of a place at a leading show.

The contention of the founders was that, besides giving that 'especial and peculiar recognition' of the native breeds, the Welsh show would also enable Welsh farmers to see once a year the best types of British breeds. Without doubt this last aspiration was realized across the whole range of livestock. In the cattle sections, Shorthorns and, though to a lesser extent, Herefords were well represented, both breeds exhibiting some of the finest animals in the kingdom, including several Royal Show winners as, in the 1908 show, the Shorthorn champion bull Chiddingstone Malcolm belonging to Sir Richard Cooper of Lichfield. The king's exhibits in the Hereford classes over the years

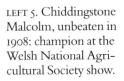

LEFT 5. Chiddingstone Malcolm, unbeaten in 1908: champion at the Welsh National Agricultural Society show.

RIGHT 6. Hereford bull Endale, belonging to Peter Coats, champion at Aberystwyth in 1907.

aroused much public interest, albeit, we have mentioned, such was the high standard of competition that his stock were often beaten, as in the 1907 show by Peter Coats's champion bull Endale.

Sheep other than Welsh Mountain and Kerry Hill likewise boasted fine specimens at the shows, as did Shropshires at the 1905 show, notably the exhibits from the far-famed Shenstone Court flock, the property of Sir Richard Cooper. One celebrated ram from this flock was Heredity, belonging to John Rees, Dolgwm, Llanybydder, which took the prize for the best ram of any age at the 1906 show. Another notable winner with his Shropshires was Alfred Tanner of Shrewsbury, whose Shrawardine Dream took first prize at both the Royal and the Welsh National in 1908. So frequently did Shropshires take the prizes over Ryelands in the early shows that the judges, following both the 1908 and 1909 shows, recommended that separate classes be introduced for each breed.

Horses and pigs of all breeds, too, were exhibited. Besides cobs and ponies, shires, hunters and hackneys were on show, with Montgomeryshire exhibitors carrying off the palm in the shire horse competitions, doubtless due to the enterprise of the Montgomeryshire Entire Horse Society. A notable shire exhibitor of these Aberystwyth years was W. F. S. Humphreys of The Gaer, Forden, Montgomeryshire, his well-known thousand-guinea horse Hendre Baronet winning the prize for the best shire stallion in 1905, 1906 and 1907. It was the case, too, that for all their ungainliness and the small number of exhibitors, pigs, as the 'Gintleman who pays the rint', were an important feature of the shows. As never before, Welsh farmers were given the chance to see the standard of excellence reached through attention to breed and judicious feeding of pigs exhibited by the likes of the duchess of Devonshire. Indeed, the 1907 show attracted nearly all the best pig champions of England.

7. Hendre Baronet, the property of W. F. S. Humphreys, The Gaer, Forden, Montgomeryshire.

Innovations were now, as throughout, to signify a developing and dynamic annual show. Early 1907 thus saw the Society's Council recommend to the general meeting of members that an auction sale of stock be held in the showyard on the second day of the forthcoming show, with the conditions of the sale to be broadly the same as those of the Royal Agricultural Society of England. Although the sale on 24 July attracted a good number of entries, there was a scarcity of buyers. If the auction at the 1908 show did witness an improvement in horses, once again it was not an overall success.

The implements section was also a prominent feature of these first six years, with some of the principal manufacturers of the kingdom occupying stands. A popular stand displaying up-to-date farm machinery, such as the Osborne self-binder so suited to hill-farming conditions, was that of Messrs M. H. Davis and Sons of Bridge Street, Aberystwyth. Among these stands of implements, too, were exhibits of various cattle feeds, flour, seeds and manures such as basic slag. That there were many more trade stands in the 1906 show than hitherto was taken by the editor of the *Journal* as a firm indication of the growing importance of the show.

Visitors to the first show in 1904 had come away disappointed at the lack of a butter and cheese section, but this was more than put to rights at the following year's show by the addition not only of the dairy section but also the butter-making demonstrations. At the Aberystwyth shows some thirty competitors entered annually for the dairy produce competition and, after their introduction in 1906, some twenty-five for the butter-making competitions. Testifying to the important contribution made to the infant society by the University College of Aberystwyth, the butter-making demonstration in 1905, for instance, was given by Miss Brown of the Dairy Department at the college and the department organized butter-making competitions on the showfield from 1906 onwards which attracted many visitors to the tent. Here was the important educational side of the show fully in operation.

The two-day show at Aberystwyth – held first at the Vicarage Field, Llanbadarn Road, and then, from 1908, at the adjacent Plascrug Meadow – with its sheds for livestock, tents, marquees, and the large ring for exhibition and jumping overlooked by a grandstand for spectators, moved the *Western*

8. The first show at Aberystwyth in 1904.

Mail reporter to write of the 1905 event: 'The appearance of the busy, convenient showyard, with its rows of shedding and canvas, is conspicuously unlike that of minor county shows, and an air of confidence pervades the whole.' Distinct programmes were put on each day. Whereas the opening day was mainly taken up with livestock judging, the second, known as 'popular day', was given over to the parade of prize-winning animals, farmers' and tradesmen's turn-outs, butter-making competitions, a butter-making demonstration and, always the highpoint of the afternoon, harness and jumping competitions, all to the accompaniment of a local band. For all the hard work and preparation of the organizers, no show could be made safety proof against things going wrong, however. During the jumping competition on the Wednesday afternoon of the 1905 show, a valuable horse belonging to Mr Wheeler, of Studley, the first performer over the jumps, dropped dead immediately after leaving the ring.

The varied attractions on the second day usually ensured a bigger gate than on the first. In 1908, for example, day two saw an estimated 10,000 people on the showground, double that of the opening day. In fact, the 1908 show's estimated total attendance of over 15,000 was the record for the six Aberystwyth shows, with those of 1904 and 1905 (the other shows for which

figures are available) in each case having attracted about 9,500 to 10,000 people. Travel to Aberystwyth by such large numbers of agriculturalists and others was greatly facilitated by the cheap-day excursions laid on by the railway companies from all the populous centres of north and south Wales. During the 1908 show twenty-three special trains arrived at Aberystwyth, comprising 224 cattle trucks and horseboxes and 100 passenger coaches, in addition to the 'ordinary' traffic. A hint of things to come was contained in applications made to the secretary, Loveden Pryse, in the months before the 1908 show from the owners of motor cars seeking permission to enter the show field. These applications were granted and the motors entered by a special gate to take up their stand by the side of the big ring remote from the grandstand.

The weather has always been a major factor in determining a show's success but this was all the more crucial in the early days at Aberystwyth and, indeed, for many years beyond, when, in the absence of a strong membership, the Society depended heavily on the gate money for its existence. Apart from the first morning of the 1905 show when the rainclouds adversely affected attendance, the weather for the six Aberystwyth shows was on the whole kind to the organizers. However, the gate at the 1909 show suffered from the vagaries of the weather for, although the actual show days themselves were sunny, the weather had been so bad for weeks before that a great many farmers were obliged to forego the pleasure of visiting the show in order to complete the hay harvest.

Attendance was also affected by the dates on which the show was held. Thus the 1907 show experienced lower gates because the event fell earlier than usual, the date having been moved back from early August to 23 and 24 July at the request of the Aberystwyth townspeople. Consequently, farmers were busy with the hay, although this earlier date would not have affected the gate quite so much as it did had it not been such an exceptionally backward season, which meant that many were busy at the hay who would, in the ordinary course, have finished some time before the show. Moreover, August was the height of the holiday season at Aberystwyth, and some of the visitors inevitably found their way to the showground. A clash, too, with other shows could seriously affect the level of entries. Such was the case with the first show in 1904, when outside entries were reduced by the fact that it clashed with the Great Yorkshire Show and was too close on the Royal Lancashire. Subsequently, efforts were made to avoid such clashes, as can be seen in the Council's decision to hold the 1908 show on 12 and 13 August, provided that date did not clash with the United Counties Agricultural Society show, in which case the dates were to be on 4 and 5 August.

If we consider the first six shows at Aberystwyth from the point of view of their meeting the expectations of the founders, the verdict has to be that the Society and its major activity, the show, had been successfully launched, but that some real problems were still to be overcome. On the positive side the initial intense local jealousies had died down considerably. Not least in winning over Welsh agriculturalists to the idea of a national society and show was the vision, persistence, thoroughness, courtesy and readiness to meet the wishes of all concerned shown by Lewes Loveden Pryse. His nice grasp of the need for inclusiveness was exemplified in his numbering among the list of ring stewards the secretary and officials of several agricultural shows up and down the Principality, a wise arrangement which helped to foster a good feeling on the part of the smaller societies towards the new national organization. Then again, as has been shown, numbers of entries were on the increase, with tenant farmers figuring strongly among the exhibitors; some of the best stock from England and Wales were exhibited; and attendance was encouraging. Although officials, members and visitors were not to be honoured by a royal visit to the showground at Aberystwyth, the king's consistent support of the Society and the prince of Wales's consent to becoming its patron in 1907 did much to boost the show's standing. In fact, at a Council meeting of 9 March 1908 it was queried whether the Society might be justified in calling itself the Royal Welsh Agricultural Society, since His Royal Highness, the prince of Wales had recently consented to become patron. The matter was then taken up by the Cardiganshire MP and close friend of the Society, Vaughan Davies, who informed the next Council meeting in mid–May that when he had approached the prince of Wales to become patron he had endeavoured to get 'Royal' added but discovered that the power to grant this rested with the Home Office, which was not prepared to sanction it until the Society was in a better financial position.

That unsound financial footing, which persisted throughout the Aberystwyth years, was the biggest of the problems hinted at earlier. It was to arouse much concern and fuel frustration as early as the Society's annual general meeting of 17 February 1906, where it was spelled out how small the Society's income was, in spite of the energetic efforts of Loveden Pryse to boost membership. The revenue account for the year ended 31 December 1905 made for doleful reading. Whereas income, made up of members' subscriptions (£365 10s. 6d.), donations (£309 17s. 6d.), show receipts (£1,078 19s. 5d.), and bank interest (£9 14s. 2d.), totalled £1,764 1s. 7d., expenditure, comprising show expenses (£666 5s. 7d.), prizes (£1,113 11s. 9d.), and other expenses of running the Society and producing its *Journal* (£196 11s. 11d.), amounted to £1,976 9s. 3d., leaving an adverse balance of £212 7s. 8d. Understandably, the Council meeting of the previous evening appealed for an increased number of annual subscribers from the present 181 members in order to generate a steady and increased income. This appeal for new members was to be but the first of many that would follow in the years ahead.

It was a frustrated and angry Loveden Pryse who suggested at the annual general meeting that every member who was interested in the Society should try and collect subscriptions and recruit new members. In his opinion, the subscription list was 'disgraceful', and he went on to inform his listeners that Sir Richard Cooper, an English gentleman who had given £700 towards the show, had written to him the previous day intimating that when he had made his donation he had assumed that landowners and wealthy men in Wales would follow his lead. Such men, however, Loveden Pryse reflected bitterly, seemed to give as little as they could, and he rubbed it in that Sir Richard had expressed his surprise at the apathy which had been shown. If, continued Loveden Pryse, it was 'a museum for the preservation of Welsh hats or something of that kind', they would give in their thousands, but they gave only £5 notes to the Society. Moreover, fifty years ago landowners had been content with receiving medals, but they now took the money prizes for the cattle and nobody, with the exception of Sir Richard Cooper, thought of giving anything back. If all the landowners and big men in Wales followed that example, he ended sourly, the Society would make far greater progress. This was an intemperate outburst, untypical of a man who had displayed patience in the face of much aggravation. It certainly roused W. Forrester Addie of Powis Castle Park, Welshpool, to voice his opinion that if show prizes amounting to £1,100 were to go on being offered by the Society then membership must be increased, but that this would not be achieved by casting aspersions on one class and another. What the Society had to do, he coun-selled, was to justify its position and appeal to those classes. Despite this keen awareness of the need to improve the Society's financial standing, no significant progress was secured during its remaining time at Aberystwyth, and for the first six years of its existence the Society practically lived from hand to mouth. Indeed, as would be the case in the deliberations in the 1950s on the wisdom of moving to a permanent site, a major consideration from 1906 in the debate over whether the show should leave Aberystwyth and become migratory was the financial repercussion of such a step.

The Society had, of course, intended from the outset, that the show should be held at Aberystwyth for the first three years. Signs of the tussle that lay ahead came in the discussion about the location of the 1907 show at the Council meeting in early August 1905; although no resolution was passed, a strong opinion was expressed in favour of a migratory exhibition. Conflicting views would again be aired at a meeting in Aberystwyth in September 1906. Those supporting the continuance of the show at Aberystwyth the following year pressed the town's exceptional rail facilities and its ideal geographical position as the most central spot for a meeting of exhibitors from north and south Wales. Advocates of a migratory show, on the other hand, argued that before the show could be a truly national one it must be moved from place to place, with centres like Swansea, Llandrindod and Cardiff being mooted. The vote went in favour of Aberystwyth as the 1907 venue, with the meeting seemingly swayed by the idea that the show be held there every year until some other town was prepared to come forward with a substantial inducement to remove it.

The spring of 1907 saw the Council decide to make enquiries as to whether Swansea, Cardiff, Llandrindod, Bangor, Wrexham or Welshpool would care to have the show in 1908. After considering invitations from Bangor, Swansea, Welshpool, Llandrindod and Aberystwyth to hold the 1908 show at their respective centres, the Council on 8 July 1907 recommended to the Society's Special General Meeting of members to be held on the showground a fortnight later that Welshpool's invitation be accepted. What seems to have swayed the Council in Welshpool's favour was the town's guarantee of money, namely £300 (£100 more than Aberystwyth pledged), and a free site. Foremost in pushing for the migratory show at the general meeting of 23 July were Sir Richard D. Green-Price, Charles Coltman Rogers of Stanage Park, Radnorshire, and W. Forrester Addie. Coltman Rogers claimed that membership was too concentrated in mid Wales. There were, he reminded the meeting, many other parts of Wales; north Wales, as he saw it, was full of 'sentiment' and south Wales was full of 'cash' and his wish was to move the show about in order to receive the sentiment of the north and the cash of the south. At stake, he insisted, was whether they were going to make the show a first-rate local one or whether they were going to give it a wider scope and magnificence by making it peripatetic or at least by holding it alternately in the north and the south. A point made by Forrester Addie was that when the Royal Agricultural Show of England remained in one place it was a failure, and only upon becoming migratory did it recover its financial position. But those at the meeting in favour of keeping the 1908 show at Aberystwyth eventually won the day, overthrowing the Council's recent recommendation. Their argument rested principally on financial grounds: that the risk involved in moving the show to Welshpool was simply too great in the face of the Society's present reserve of just £300. They argued that there was insufficient capital to undertake the risk of the increased expenses that would be incurred in any change from Aberystwyth. After all, the general manager, Lewes Loveden Pryse, did all the work practically for nothing, indeed he may well have been out of pocket. Moving the show would mean that the Society would have to pay its manager or get someone else, for they could not expect someone to work for nothing at a place away from home.

That there was point to this was to be corroborated by the secretary of the Royal Counties Agricultural Society in a letter from his Basingstoke office to Loveden Pryse in August 1907:

I should be very glad if my own Show was held at one centre; as a great part of my time is spent in travelling about, not only to make the arrangements for individual Shows, sometimes 50 miles away, but to look ahead for two or three years and make arrangements for future meetings, and this ought to be taken into account in fixing the remuneration of the secretary.

Furthermore, argued Aberystwyth's supporters, the expense of removing the fixtures and of fitting up a new showyard would be great. Nor, in their view, would the Society's national character, an important consideration, be eroded by the show's remaining a fixture at Aberystwyth. Indeed, it would be well served; Aberystwyth was a place where north and south could meet, whereas if removed to north Wales the show would become merely a north Wales one and, likewise, just a south Wales show if held in that region. (To an extent, we shall discover, these last prognostications were to be proved correct.) The vote was about three to one in favour of holding the show at Aberystwyth in 1908.

The issue of the show's location was so frequently talked about within Wales and the Council was so concerned to reach the right decision that it persuaded the Society's members to vote on the location of the 1909 show. Aberystwyth emerged the clear favourite, with 145 members voting for it as against 49 for places elsewhere. However, if we look at the votes of those members residing outside the Union of Aberystwyth the margin supporting Aberystwyth was significantly narrower, 61 voting for that town and 44 for other centres. At the following annual general meeting in February 1909, pro-Aberystwyth people like Vaughan Davies, MP, A. J. Hughes, the town clerk, and Lewes Loveden Pryse once again pushed the view that the Society's financial position was not strong enough to justify any change in the location of the show. Nevertheless, they stressed that they would gladly support a change if any of the towns wishing to host the show came forward with a substantial guarantee against any possible loss which might be incurred through its removal from Aberystwyth. And here the point was made that if it was necessary to obtain a guarantee of £1,000 or more for the National Eisteddfod it was equally necessary to have a like fund for the show. Sensitive to a growing feeling that the Council was prejudiced in favour of Aberystwyth, and that there was a certain selfishness to Aberystwyth's supporting the retention of the show there, Professor Bryner Jones of Aberystwyth recommended that the Council make it clear that this was not the case but that, rather, they the Council were willing to move provided other places would give the necessary guarantee. Laughter in the meeting greeted the quip of an Aberystwyth and Loveden Pryse supporter, C. M. Williams, that even if the show was located in paradise, there would be grumbling and fault-finding. Aberystwyth once again won the day, a decision sufficiently unacceptable to outspoken critic of Aberystwyth, George R. Pryse, to prompt his on-the-spot resignation from the Council as a representative for Cardiganshire no less.

At the last Aberystwyth show – in 1909 – much discussion took place as to the future location. In an editorial for 6 August 1909, the *Cambrian News*, now in more positive tone, pointed out that the question was rendered especially difficult because of the configuration of Wales, for a show held in Radnorshire, Anglesey, Pembrokeshire, Denbighshire, Glamorgan, Flintshire or Caernarfonshire could not be expected to secure attendance from the rest of Wales and so would be only nominally national. Aberystwyth, it was argued, was fortunately situated as a centre for the rest of Wales. It also had plenty of level ground available. It was important also that the new manager, Professor Bryner Jones, head of the Department of Agriculture at the College at Aberystwyth, and who had taken over in 1909 after the resignation of Lewes Loveden Pryse through ill health, possessed the necessary knowledge, skill and time. Indeed, the editor was to suggest that: 'In a sense, the Welsh National Agricultural Show may be looked upon as a sort of institution identified with the three Welsh University Colleges.' Without Bryner Jones's services, he ruminated with just a hint of blackmail, a permanent secretary would have to be appointed.

The Society changed its mind regarding the future location of the show in the months following the 1909 exhibition. Seemingly crucial in persuading the Council to uproot the show from Aberystwyth was the balance sheet for the year 1909: whereas total income amounted to £1,886 13s. 3d. (£727 17s. 1d. from subscriptions and donations plus £1,158 16s. 2d. from show receipts), expenditure totalled £2,158 11s. 4d. (£1,161 6s. 2d. for prizes, £696 10s. 5d. for show expenses and £299 14s. 9d. for miscellaneous expenses), which left a deficit on the year's workings of nearly £272. Here taking its toll was a decrease of £205 in the show's gate receipts and of £235 in subscriptions. The chairman of the Council, Sir Edward Pryse of Gogerddan, said he believed that the decision at its last meeting not to make the show a migratory one had prompted several gentlemen to withdraw their subscriptions. Accordingly, the Council meeting of 23 November 1909 called for approaches to be made to the town authorities of Swansea, Newport, Wrexham and Bangor as to the feasibility of holding the 1910 show at any of those centres. What ultimately emerged, however, was that the Council at its meeting in February 1910 recommended acceptance of the invitation of the Carmarthenshire Agricultural Society to hold the 1910 show at Llanelli. The Carmarthenshire

Society undertook to provide a suitable ground of about twenty-five acres with a permanent grandstand and horse ring, together with a subscription of £200. It was recognized that Carmarthenshire's offer was a very generous one, both in its willingness to cancel its own local show and in the readiness of its officials to complete the local arrangements, and the Council went on to stipulate that any profit from the show remaining after the present deficit was cleared should be handed over to the Carmarthenshire Society. At long last the Council had come round to realizing the advantage of holding its show in a populous centre like Llanelli, and that if the show was moved from Aberystwyth more members would join the Society. The decision to move was welcomed by the *Western Mail*, which concluded its editorial thus: 'There is no reason why a flourishing Welsh Agricultural Society should not be built up, but it must appeal to a much wider circle than it has yet reached.' The Society had embarked on one of its many big gambles; the wisdom of so doing only time would tell.

CHAPTER THREE

Far More Than Just a Show

So concerned were the Society's founders that their new organization was perceived by the public at large as having been set up with the sole purpose of holding a big annual show that as early as the second number of their *Journal*, produced in January 1905, an article was devoted to spelling out the Society's principal aim. Briefly, this was 'to assist agriculture generally throughout the Principality in every way in its power'. It was immediately conceded that the show was one of the Society's key activities, furnishing as it did such a splendid object lesson in both breeding standards and the availability of new implements to those farmers who lacked the means to attend the big shows further afield. But there were a great many other ways, the writer (most certainly either Lewes Loveden Pryse or D. D. Williams) urged, in which the Society could assist the farmer and the agricultural interest in Wales.

One such cause that had already been taken up by the Society was the terrible scourge of sheep-worrying by dogs in Wales, with more than 4,000 sheep and lambs having been worried in Caernarfonshire alone in 1904. At the Council meeting of 17 January 1905 held at Llandrindod it was decided to write to the president of the Board of Agriculture, drawing his attention to the necessity that every dog which had been granted an exemption should wear a collar bearing a distinguishing metal tab, thus making the detection of unlicensed dogs easier for the authorities. Another matter the writer recognized as having already received the attention of the Council appertained to the Fertilizers and Feeding Stuffs Act (1893), the same meeting at Llandrindod having resolved to write to various county councils in Wales asking them further to enforce the said Act by appointing officials to take samples. Such action was urgently needed, for, as things stood, after purchasing artificial manure or feeding cake the farmer, in order to satisfy himself that it was up to the standard guaranteed, had to go through a vast amount of trouble and formality before he was able to get an analysis.

All this won the recognition of the *Western Mail*'s special correspondent on 2 August 1905:

The Society not only aims at doing good work in the show line, but aspires to still more beneficial spheres of labour. It is a splendid medium between the farmers of Wales and the Board of Agriculture, and representations made to the Board with regard to sheep-worrying by dogs, compulsory sheep-dipping etc. have already had good effect. It has also communicated with the county councils in Wales and Monmouthshire calling their attention to the necessity of putting the Fertilisers and Feeding Stuffs Act, 1893, into more effective operation, and in many ways has identified itself with agricultural movements of the day. A journal is also issued quarterly, and it is hoped that gradually farmers will utilise the resources of the society as the medium whereby they can focus their efforts for united action.

The annual general meeting of the Society on 17 February 1906 was informed by its chairman that the Council had been negotiating with the Board of Agriculture on the matter of sheep-dipping areas. In some cases they had already been successful in having the boundaries altered for the benefit of large flock masters, and from summer 1906 a new Compulsory Sheep Dipping Order for the whole of Wales and Monmouthshire would come into force. As the *Journal* for April 1906 commented, the measure, while unwelcome to some farmers and viewed as a great hardship by those who had no scab in their flocks, would be welcome to most Welsh farmers because it was far better to have one universal dipping and see for certain whether it had any appreciable effect in stamping out or lessening scab than to try it in one or two small districts, as had been done in the previous year, when the dipped sheep then went out of the area and mixed with other scabby sheep. As a result, there was no way of knowing whether the dipping had been ineffectual or whether the sheep had been reinfected.

As well as reporting further on the efforts being made in the matter of sheep scab, the Council meeting of 21 February 1907 also detailed its endeavours to obtain the alteration of the existing law relating to the taxing of farmers' carts. In fact, the issue of officers of the Inland Revenue making farmers take out licences for light dog carts had been under consideration for some time, and the stance taken by the Society in defending three of its members the previous December had been that a light dog cart was essential for the purposes of his trade to a farmer who went in for breeding light horses, and that he was not liable for a licence provided he did not use the trap for purposes other than his trade. The Society also urged that driving a light dog cart in the show ring, for which many had been summonsed or made to take out licences, was using it for the purposes of his trade since a farmer took it there the better to show off his horse in the hope of selling it. The magistrates, on that occasion, decided that provided the farmer's name was painted on the cart in proper-sized letters he did not have to procure a licence. However, since so many farmers, fearful of receiving a summons, had already obtained licences, the Society had proceeded to communicate with the Treasury, and the Council held out grounds for hope that the farmers would in due course be relieved of the problem.

In its Aberystwyth years the Society supported other campaigns helpful to the Welsh farmer. At its meeting on 16 February 1906 Council resolved to endorse the Welsh Pony and Cob Society's action in endeavouring to bring about legislation over the matter of clearing the hills and commons of Wales of undesirable stallions. A deputation to the minister of agriculture took place in July 1906 with the Welsh National Agricultural Society represented by Lord Carrington, Loveden Pryse and Vaughan Davies. In a later Council meeting of 22 June 1907 members approved the views expressed by that deputation on a number of matters, including:

1) the permissive legislation called for to clear the hills of undesirable stallions would protect a majority of Commoners who were endeavouring to maintain their pure breeds of Mountain Ponies against a minority of Commoners who frustrated those attempts by neglecting to exercise any supervision over their pony breeding stock on the hills and commons;
2) that it would tend to facilitate the efforts that were being made to induce the authorities at Washington (USA) to lift the heavy import duties levied at their ports on Welsh Ponies, and to grant them the same privileges extended to other registered breeds, as, for example, Thoroughbreds, Cleveland Bays, Polo Ponies, Hackneys, Shires and Shetlands;
3) that under the protection of such a law the value of Welsh Mountain Breeds would tend to be greatly enhanced and
4) that owing to the permissive character of such proposed legislation it could only be put into operation on such common lands on the clearly expressed wish of the majority of Commoners.

Real results were soon to be seen in the announcement by the secretary of the Welsh Pony and Cob Society at its meeting on 16 September 1907 that he had received news that the United States government had now agreed to recognize the Welsh Pony and Cob Society so that, if the information was correct, their ponies would in future enter the country free. Furthermore, at the annual general meeting of the Welsh National Agricultural Society on 10 February 1909 it was reported that a bill had been introduced to Parliament with reference to clearing the hills and commons of undesirable sires.

At the Council meeting of 16 February 1906 Vaughan Davies was appointed to represent the Welsh National Agricultural Society in conjunction with a deputation from the Welsh Black Cattle Society, which included prominent north Wales breeder, R. M. Greaves, to the president of the Board of Agriculture for the purpose of opposing the admission of Canadian cattle into the country. Reflecting the close interpersonal ties between the two societies, Greaves was also to be a representative of the Welsh National Agricultural Society. Later, at a Council meeting on 6 October 1908 it was resolved that the government should be urged not to relax the prohibition of importation of foreign store livestock, and that a copy of the resolution should be sent to the president of the Board of Agriculture and every Welsh member of Parliament.

Yet another manifestation of the Society's early involvement in areas beyond the show itself is reflected in the following resolution initially passed by the Pembrokeshire Farmers' Club and then later considered and endorsed by the Council on 22 June 1907, with the secretary being instructed to forward a copy to the president of the Board of Trade and to each of the Welsh county members of Parliament: 'That it is the opinion of this meeting that the permission given by the Government to the Army and Navy Contractors to supply Foreign, Frozen, and Tinned Meat to the Forces is detrimental to the Home Grown Produce, and a heavy tax on the farming community.' Reflecting the Society's influence within the world of farming in the United Kingdom was its representation on outside bodies; for instance, in 1907 delegates were sent to the Royal Agricultural Society of England Committee on Tuberculosis in Cattle and the National Poultry Conference, while illness prevented a delegate from attending the February conference of representatives of Agricultural and Breed societies.

An activity of the Society undertaken from the outset and which was to continue, despite interruptions, throughout its first hundred years was the publication of its *Journal*. Introducing the first number of October 1904, the editor wrote that the Society felt confident that it would fulfil a long-felt need in Wales: 'The aims and objects of the Periodical are mainly to record the work of the Society, and also to publish articles in Welsh, as well as English, dealing with the breeding and improvement of livestock, others on agriculture in its scientific aspects, etc.' Optimistically, the editor urged members to pass the first number on to a friend or a neighbour and at the same time induce them to join the Society. A reminder was given that tenant farmers could join the Society on paying 10s. 6d. as an annual subscription and obtain all the privileges of membership, including a free copy of the *Journal*. Stress was laid on the fact that no publication of its kind had been issued before in Wales, and that all the agricultural literature hitherto available to Welsh farmers had come from over the border.

Some eighteen issues were to be published between October 1904 and January 1910, an impressive achievement, but thereafter its publication was discontinued doubtless as an economy measure in the face of financial difficulties. In fact, no further volumes were published until 1923. The eighteen numbers mentioned make for some absorbing reading, especially in the early ones, ranging over articles on 'Welsh Cobs and Ponies' by Richard Green-Price, 'Welsh Mountain Sheep' by Marshall Dugdale, 'Shire Horses' by Edward Green, 'Kerry Hill Sheep' by T. Halford, 'Fowls on the Farm' by Bessie Brown, and 'The Welsh Black Cattle Society and its future work' by J. B. Owen. Articles written in Welsh included 'Gwartheg Cymreig' (on the cattle of Wales) by Thomas Roberts of Aber near Bangor, 'Gwrtaith a Gwrteithio Calch' (on organic and artificial fertilizers) by D. D. Williams, 'Porthiant a Phorthi Anifeiliaid (on feeds and feeding) by John Roberts of Towyn, and two anonymous articles respectively on ensilage, 'Y Modd i Drin Gwair Glas', and on fertilizers with particular reference to potatoes, 'Gwrtaith Celfyddydol i Bytatws'.

Some criticism of the *Journal* in its Aberystwyth years is merited. While it is true that agricultural education and research was in its infancy, surprisingly few reports of experiments carried out on the College farms were published even though the College departments were always represented on the Society's Council. The earlier numbers carried much valuable farming information, but it does appear that over time its quality deteriorated. Yet, tribute must be paid to the committee appointed in 1905 for overseeing the *Journal*, namely D. D. Williams, its likely editor, Coltman Rogers and John Roberts.

PART TWO

In Search of a Truly National Identity

THE FIRST MIGRATORY PHASE, 1910–1939

CHAPTER FOUR

Years of Anxiety

W̲e have already seen that it was soon felt that the practice of holding the show in one fixed place was not altogether in keeping with the 'national' character envisaged for the Society by its founders and that it tended to localize interest taken in its activities. As a result, the show became a migratory one from 1910 onwards, meeting at centres in north and south Wales alternately as far as possible. In the early years from 1910 to 1913 the show continued as a two-day event, but from 1914 onwards it was held over three days.

At the beginning of its itinerant phase, the Society itself underwent changes to its constitution. In future, the Council was to consist of forty-two elected members: thirty-six were to be representatives of the twelve Welsh counties, each having three apiece, and a further three were to represent Monmouthshire. Each county elected its three members. Moreover, reflecting the importance of the Welsh University Colleges in the affairs of the Society, each of the three colleges was to have one representative, thereby making a total elected membership of forty-two. Members were to retire every three years, but were eligible for re-election. The Council was to elect a president each year – in effect from the county hosting the annual show – trustees, vice-presidents, and any other officer required, who were to be ex officio members of the Council during their term of office. Importantly, the Council was to have full powers to choose the venue of the show and the location of the headquarters of the Society. In so far as the siting was concerned – excluding the seven years 1915 to 1921 when the First World War and its aftermath ruled out any exhibition, and again in 1938 when the Royal Welsh merged with the Royal Show at Cardiff – between 1910 and 1939 some ten were held in the north, ten in the south, and two in mid Wales. Throughout these years the Society's business was run from offices at Wrexham, while Council meetings and standing committee meetings, apart from those held in the various show weeks on the showground, took place in Shrewsbury.

The years immediately following the switch to a migratory show witnessed further interest in the affairs of the Society by the royal family. We have seen that the prince of Wales became patron in 1907 and presented a silver challenge cup to be competed for annually by the best Welsh cob. In early 1911 the Society was informed that the new king, George V, had consented to become patron of the Society, the prince of Wales, vice-patron, and that the prince would present a challenge cup. The cup presented in 1907 would in future be known as the George Prince of Wales Challenge Cup. Always seeking to promote native livestock, the Council recommended that the new cup, styled the Edward Prince of Wales Challenge Cup, should be offered for the best group of four Welsh Black cattle (the property of the same exhibitor), with a minimum one male and one female in each group.

Such valuable royal patronage would have helped to offset some of the problems besetting the Society on its departure from Aberystwyth. For, after the successive shows at Llanelli (1910), Welshpool (1911), Swansea (1912), Porthmadog (1913) and Newport (1914), no great turnaround in its fortunes had been achieved. The problem of low membership still persisted and pre-occupied the Society's officials. Notwithstanding the confident assertion of the latter at the Swansea show in August 1912 that the result of the show 'is to set the society on its legs and give it an assured position for the future', the president went on to say that a larger number of members was required. A lack of members meant that at the time of its suspending activities for the duration of the First World War and beyond, the Society's finances were shaky. It just took one unsuccessful show as far as the 'gate' was concerned to bring problems. Certainly, the disappointing financial result accruing from the 1914 Newport show prompted the treasurer, Arthur Jones, to ask the Council in February 1915 to prepare for an overdraft of £450 at the bank to meet the Society's liabilities. When Colonel David Davies of Llandinam, MP, and two others volunteered to act as guarantors to the Society, all members were asked to become sureties in the sum of £5 to these principal guarantors. In this crisis, Colonel Davies demonstrated yet again his unstinting devotion to the Society's welfare, though his willingness to become a guarantor was contingent on the appointment of a secretary from the first Council meeting after 29 September 1915 and that whomsoever was chosen should have his personal approval.

Such a low ebb in the Society's funds and membership explains the delay in holding its first show after the war until 1922, a deferment which contrasted with the prompt restarting of other shows such as those of the Royal Agricultural Society of England and the Bath and West of England Society. The Welsh National Agricultural Society was informed by their treasurer at the Shrewsbury Council meeting of 30 March 1921 that £70 was owing to the bank, that certain accounts had not been settled and that several cheques for things had not been cleared. So hard pressed was the Society that it was resolved that David Davies, MP, who had been elected chairman of the meeting, should write to the prizewinners inviting them, in view of the financial position, to forgo the prize money due to them. The continuing unsatisfactory level of membership was still the Achilles heel of the Society and prompted its chairman to make an appeal for support at the start of 1921. In the light of this insecure financial position, it was decided that no show would be held in 1921 and, indeed, it was resolved that the next Council meeting be held as early as possible in October or November 1921 when it was hoped that information and estimates would be forthcoming which would enable it to decide whether it would be possible to hold a show in 1922. The Council meeting for 25 November 1921, with David Davies, MP, in the chair, resolved to proceed. At this same meeting the chairman reported that the current membership was 450, and that £480 6s. 6d. had already been received in subscriptions, with £303 to come from those who had promised to join. He hoped that before the following summer the sum to the credit of the Society at the bank would amount to £1,000. Of huge significance for the Society's future was the fact that at this November meeting David Davies was appointed to the office of chairman of the Council, an office he was to hold until his death in 1944. His successor as chairman, Sir C. Bryner Jones, recalled the importance of this juncture in the Society's history in a lecture he delivered to the Society on 12 July 1954: 'But in 1921 – mainly through the interest and enthusiasm of Mr David Davies (later Lord Davies of Llandinam) – it was decided to restart the Society and in 1922, a successful show was held in Wrexham, thereby starting a new period in the history of the Society.'

So disastrous was the weather on the first day of the Wrexham show in late July 1922 that David Davies suggested to the Council, which met on the showground, that in view of the Society's poor financial position at the close of the show, it would be well to postpone further arrangements for future shows until a firm financial statement relating to the present event could be placed before the Council, a course of action which was adopted. His gloomy expectations were happily dispelled, however, by the news that the Society had emerged from the Wrexham show with the largest balance it had ever realized. The spirits of organizers and members alike were further lifted in late 1922 when they were informed that the king had granted the Society the use of the prefix 'Royal' in its title and had commended that henceforth it be known as 'The Royal Welsh Agricultural Society'.

At this juncture, the Society could have justifiably felt some quiet satisfaction. The Wrexham show of 1922 had placed it on a sound financial basis by making a profit of over £2,150, and the following month witnessed a marked increase in membership. As a further token of royal patronage, upon its resuscitation in 1922, the prince of Wales honoured it by agreeing to become its president, a position he held for the second year running in 1923, and from 1924 he became permanent honorary president of the Society. It was

9. The Wrexham show, 1922.

owing to his inspiration, not least his letter of appeal for support, that the membership of the Society, which had never exceeded 300 in pre-war days and stood at the record figure of 576 for the Wrexham show, had by the Welshpool show of late July 1923 grown to 1,040. The Society was also at this time to acquire new officials of outstanding calibre. Arthur Evans of Bron-wylfa, Wrexham, who had years of experience as a breeder of hackneys and a supporter of the Welsh pony and cob, had become the honorary director in 1921 and would continue in that post until his untimely death from appendicitis in August 1926, less than a fortnight after successfully seeing through the Bangor show in July. It was due to his energy and management skills that the three-day show held at Wrexham assumed unprecedented proportions, in terms of exhibits and revenue, thereby providing the means by which the Society was placed on a financial basis that gave promise of stability and an expanding range of activity. Under his direction, which combined a wide experience of shows, business acumen, a reputation for straight dealing, and courtesy and tact, the later shows at Welshpool (1923), Bridgend (1924), Carmarthen (1925) and Bangor (1926) were admirably conducted and achieved notable success.

Of even greater importance for the long-term viability of the Society was the emerging influence of Colonel David Davies of Llandinam, MP, one of the most influential stockbreeders in the country. He had been the driving force behind its resuscitation in 1921 in response to approaches made to him in late 1920, at a time when the Society had no officers and no funds, by some anxious to revive the Society, including C. Bryner Jones and Welsh cob and pony breeder Tom Jones Evans. We have seen that David Davies was elected chairman of the Council in 1921. For his services to the Society and many other public bodies in Wales he was elevated to the peerage in 1932. No one was as influential in shaping the development of the Society during the 1920s and 1930s. Rarely did he miss a Council meeting and, if at times he was prone to be a little hasty and somewhat irascible, his sound judgement, especially in matters of finance, vision, strength of resolve and financial generosity, were crucial in enabling the Society to weather some severe storms arising from the agricultural depression of the inter-war years. Professor Moore-Colyer indicates that, although among Welsh agriculturalists it was the hill farmer who endured the hardest struggle throughout the 1920s and 1930s, at the depth of the depression between 1929 and 1932 many Welsh farmers, including those of the relatively more prosperous lowland areas concentrating on dairying, were to suffer significant financial loss. Sir C. Bryner Jones, in the

10. Lord Davies, chairman of the Council from 1921 to 1944.

same 1954 lecture referred to earlier, concluded thus: 'It is impossible to measure the late Lord Davies's service to the Society and to the show. It is to him more than to anyone else that their success and growth in the years after the First World War are attributable, not only because of his generosity but also because of his personal interest in all of the Society's activities.'

David Davies's conviction that if the Society was to be assured of survival it must secure a strong financial position led to an early alteration in the way the Society presented its accounts. By September 1923 he had divided the accounts into two, the management account, pertaining to the Society's income and expenditure, and the show account, which was primarily concerned with the financial results of the annual show. In order to assume a secure financial footing it was essential to build up a reserve fund which could best be achieved by each year investing the profit on the management account. As early as the Council meeting of November 1926 the honorary treasurer drew particular attention to the satisfactory progress that had been made in the Society's finances in connection with the establishment of a reserve fund; investments at that time totalled approximately £5,000.

A further alteration in the Society's organization came in 1924 in response to C. Bryner Jones's complaint that the Council had so little say in the management of the Society, meeting as it did just once a year. Since this caused considerable comment and was agreed not to be in the best interests of the

Society, it was decided to hold quarterly Council meetings. Not only were Council members thereby encouraged to show more interest in the Society, but an attempt was also made in late 1926 to stimulate interest in its activities by inserting reports of Council and committee meetings in the press. A further crucial development in the democratization of the Society came from 1924 onwards when county meetings were invited to submit to the Council valuable recommendations and suggestions about the running of the Society and the show. That the value of such county meetings was keenly appreciated by the Council and its chairman, David Davies, is seen in their instructing the secretary in July 1927 to attend for the first time each of the forthcoming meetings to endeavour to establish, and maintain, personal contact between the various counties and the Society's headquarters. Colonel Davies contended that this 'personal touch' had previously been 'sadly lacking', and he believed that subsequent visits paid by the secretary that year had done a lot of good. Equal enthusiasm for county meetings came from the localities: the Breconshire county meeting in 1937 called upon the Society's organizers to allow for more than one meeting a year in order to foster sustained interest in the Society and its work.

Improvement also came with the reorganization of the Society's office staff in 1927, necessitated by the much-lamented death of Arthur Evans. A general secretary to the Society was to be appointed at an annual salary of £500, together with reasonable travel and out-of-pocket expenses; the further post of accountant and assistant secretary was to be offered to Walter Williams, who had indefatigably served as secretary hitherto. Shouldering the administration of the Society's entire affairs as the chief executive officer, the new general secretary was to be accountable solely to the Council. From among 160 candidates, the final choice of the selection committee in early 1927 went in favour of Captain T. A. Howson.

Howson came in at a testing time. The Society was seriously feeling the knock-on effects of the agricultural depression and it was urgently seeking ways to economize. Part of the problem bedevilling the Society was the high level of unpaid membership subscriptions; by April 1929, of a total membership of 1,062, only 488 members had paid, the remaining 574 being in arrears. Equally serious was the low level of membership itself. In late May 1929 the Finance and Executive Committee debated a motion that the services of one member of the office staff be given over exclusively to drumming up membership, only to reject it on the grounds that every possible effort was already being made in that direction. Of course, the Royal Welsh was not alone in facing membership difficulties during the depression; agricultural societies in general were experiencing difficulty in maintaining membership and collecting subscriptions during the grim year 1932. During 1930 David

Davies had urged fellow Council members that expenditure on the show account for 1931 should be reduced wherever possible, provided that the show did not suffer materially as an attraction to the public and that its value as an educational exhibition was not curtailed. One way in which this might be done was by restricting expenditure on prize money and judges' fees and expenses. In similar vein, in December 1931 the chairman urged that either the forestry or poultry sections – money losers at previous shows – should be omitted from the 1932 event. The outcome was that the forestry section was excluded and would remain so for a number of years to come. Even though it was a pet project of David Davies, first implemented in 1924, consideration was given in early 1932 to the omission from that year's show of the Inter-County Group and Breed Competitions; while the specifically Welsh breed competitions under Scheme A were retained partly through the personal bounty of David Davies, Scheme B was dropped. These economies prompted Professor R. G. White of the Department of Agriculture of the University College of North Wales at Bangor to declare that the Society had reached, if not actually passed, the minimum of activities which a national show should undertake and to call for a bolder policy for the Aberystwyth show in 1933. His plea was to no avail, however; Scheme B of the County Group and Breed Competition was not reinstated in the 1933 and 1934 shows and the forestry exhibition was only revived at the Abergele show of 1936.

Such a rigorous curtailment of expenditure achieved the desired effect of maintaining the Society on a reasonably good financial footing during the critical years of the early 1930s. Net surpluses on the management and show accounts were achieved in 1931, 1933 and 1934, while the net loss of £241 sustained for 1932 was no great disaster. This meant that the society's reserve fund continued to grow, reaching the psychologically satisfying total of £10,000 by the close of 1934. Small wonder that Lord Davies in his *Report on the General Working of the Society* for 1934 was in good spirits: he recalled that on restarting after the war the Society's total assets amounted to a mere £163 16s. 9d, that in the interval it had weathered a protracted period of economic depression, and that, alarmingly, it had sustained a loss of some £2,000 on its 1927 show. Yet he felt that the Society had every right to feel proud of its achievement in building up this total. Indeed, such a satisfactory position was to lead the prudent chairman at the 1934 show to express the hope that the fund would soon be sufficiently substantial to enable some reduction in the amount of the donation or guarantee required from smaller centres – towns like Corwen, Lampeter and Llandeilo – anxious to host the Royal Welsh Show. Requests for such a relaxation were forthcoming from various county meetings for 1934, but the Society's ruling body deemed that the time was not yet opportune for this reduction. They were to be proved correct.

Despite Lord Davies's comment at the 1937 Monmouth show that after sixteen years they had built up a reserve fund that would enable them to weather any storm, a sharp reminder of the Society's financial vulnerability came with the deficits incurred at the successive Abergele and Monmouth shows of 1936 and 1937. Deplorable weather conditions played a big part in this unhappy state of affairs. Liquidation of the total deficit over the two years of £1,748 meant that the reserve fund had to be raided, and was a stark reminder of how quickly a succession of unfavourable years might erode the Society's resources. In sombre mood, Lord Davies was at pains to remind members in late 1937 that until the Society succeeded in strengthening its financial position even further it continually ran the risk of being forced to suspend its operations. He fairly claimed that since the relaunch in 1922 every effort had been made to build up the Society's resources, but the serious reverses of the previous two years left no option but to make serious inroads into the reserve fund. As anticipated, further financial loss would inevitably be sustained in 1938 because the Society's own show had been suspended on the occasion of the visit of the Royal Agricultural Society of England Show to Cardiff in the summer of that year. E. Verley Merchant was to reflect sourly that, while it was all very well for people to applaud the Royal Welsh Society for this act of grace, it meant that there was a heavy financial burden to carry. All these setbacks meant that over the three years 1936–8 net losses amounting to £2,085 0s. 5d. were sustained, which in turn saw the Society's assets, worth £11,319 11s. 2d. at the close of the 1935 financial year, fall to £9,234 10s. 9d. by autumn 1938. The situation was indeed perilous, and stemmed principally from the fact that revenue from membership subscriptions was insufficient to meet administrative and other overhead expenses during a normal year. Therefore, the Society was in the precarious position of having to rely entirely on the success of its shows not only to consolidate a reserve fund, but also to balance its annual budget.

Readers will need no reminder that, for all the careful preparation and hard work beforehand, the success or failure of an agricultural show largely depends upon an uncontrollable factor, the weather. Should successive shows be hit by adverse weather, resources would be quickly dissipated. The Society was clearly in need of a regular and assured income to enable it both to carry on managing its ongoing affairs and to build up a sizeable reserve fund which would place it upon a firm foundation, and this income could only be provided by annual subscriptions.

Faced with the crisis that had blown up in the years 1936–8, in December 1937 the Society appointed a sub-committee to consider ways of increasing the reserve fund and membership roll. Meeting on 1 February 1938, it was informed that the current membership was 1,057 and the reserve fund totalled

£9,603. By comparison, the approximate memberships and reserves of the other big agricultural societies were as follows:

SOCIETY	MEMBERSHIP	RESERVES
Royal Agricultural Society of England	9,000	£231,700
Highland and Agricultural Society of Scotland	10,000	£184,300
Yorkshire Agricultural Society	2,949	£41,800
Royal Lancashire Agricultural Society	2,986	£22,450
Bath and West and Southern Counties Society	1,072	£18,579
Three Counties Agricultural Society	1,295	£10,124

Clearly the Royal Welsh was a small player and in terms of resources – though it is emphasized not in quality of its annual show – had a long way to go to match its rivals, a circumstance that could, of course, be partly explained by its late arrival on the scene. But only partly, for there is no mistaking the slowness of Welsh agriculturalists to respond to frequent membership appeals. In an attempt to turn the tide, the following recommendations were made to the Council in 1938: (1) that existing members should do their best to recruit additional ones; (2) that closer cooperation be forged between the Society, the National Farmers' Union, the Council of Agriculture for Wales, the various county and local agricultural societies, the Country Landowners' Association and such bodies, all of which were then working more or less independently in a common cause; (3) that the Council members in each county be requested to meet and devise means of increasing membership in their county; (4) that advantage be taken of Daniel Daniel's offer to approach prominent persons in south Wales and Monmouthshire to urge them to become members or subscribe to the Society's reserve fund, and that Lord Mostyn be asked to do likewise in north Wales and J. Morgan to take similar action in London; and (5) that efforts be made to persuade ten friends of the Society to subscribe £50 each to the reserve fund, subject to the Council doubling the membership within twelve months from a specified date.

The immediate response to the implementation of these recommendations was not heartening. Membership by late 1938 had risen from the 1,057 of February to just 1,133 (some 237 of whom were resident in England), although more members than usual were enrolled in 1939. Set against this last encouraging development, however, was the dismal weather on the last two days of the Caernarfon show of that year, which quickly dashed any hope of making up for the losses suffered in previous years. There is no avoiding the plain fact that during the period to the outbreak of the Second World War the Society deserved rather more support than it got from Welsh agriculturalists. Of course, one of the main difficulties in increasing the permanent membership of a society which held migratory shows arose from the fact that so many farmers joined only for the year in which the show was held in their locality in order to reap the benefits which such membership conferred and then withdrew their support. Continued rather than spasmodic membership was urgently required.

At the same time the apathy of the general public towards the venture can in part be blamed on the Society itself. Many had long felt that it should move its headquarters from Wrexham to a centre nearer the populous parts of Wales. Certainly a move had been mooted as a possibility in 1927 when Cardiff, Aberystwyth, Shrewsbury and Llandrindod Wells had been suggested as suitable alternative venues, but the state of the finances in early 1929 put paid to any further consideration of the question. Added to the unsuitability of Wrexham as its headquarters was the inconvenient circumstance of the Society's banking facilities being located in distant Aberystwyth.

For all its shaky financial foundations, the Society and its show nevertheless enjoyed steady development and expansion in the inter-war years. In part this can be attributed to the efficiency and dedication of its officials. We have already noted the significant roles played by David Davies and Arthur Evans. In passing we also mentioned the appointment of Captain T. A. Howson of Gresford in 1927. After serving with distinction in the Royal Artillery during the First World War, Howson had settled down to supporting Welsh breeds and publicizing their merits in the columns of the *Livestock Journal*. Small wonder that he was the unanimous choice for the post of secretary in May 1927, an office he discharged with such zeal, understanding and efficiency that, for all his modest and retiring disposition, he quickly won and retained the respect and liking of those with whom he had dealings. Above all, it was his close contact with the Welsh breed societies and his regular contributions in the press which promoted the development of the Royal Welsh Show and secured its growing reputation within the country at large. He was to organize his last, brilliantly successful show at Carmarthen in 1947, before retiring the following year after twenty-one years in harness.

11. Captain T. A. Howson, secretary of the Society, 1927 to 1948.

Another stalwart in these years was Reuben Haigh of Gardden Hall, Ruabon, who served the Society unstintingly as honorary director over many years. First acting as assistant honorary director to Arthur Evans, he became honorary director in 1928 and again in 1930, and, enjoying the patronage of David Davies, he was appointed permanent honorary director from 1932. The chairman believed that appointing a permanent holder to this post would improve the show's organization by introducing an element of continuity in show management, a course adopted by other leading agricultural societies.

Nothing but praise was heaped upon him by his chairman, who, on hearing in late 1937 of his intention to resign because of the pressure of business commitments, commented sombrely that his departure would amount 'almost to a catastrophe' and expressed the hope that he might yet reconsider his decision. He duly obliged and remained in post until 1950, only to die shortly afterwards in March 1951. His business experience, and also his devotion to duty, his modest, friendly, cheerful and approachable disposition and his unfailing courtesy and tact and, when necessary, firmness, ensured that the shows which he organized proceeded smoothly. Lord Davies went so far as to describe him in 1939 as 'the cornerstone in our edifice' and that under him 'the standing of the society had increased enormously'. Nor must we overlook the dedicated service to the Society of Arthur Jones in his capacity as honorary treasurer from the Society's inception in 1904 to the time of his death in 1930. He had been sometime manager of the Midland Bank at Aberystwyth and was to be succeeded as honorary treasurer by R. H. Thomas, then manager of the same bank. Continuity was thus assured.

Such was the expansion of the Society that by autumn 1936 the general secretary and his office staff were feeling the strain, and, with the loss of an experienced and reliable member the previous February, additional assistance became a matter of urgency. Details of the duties that he and his office staff had to perform were outlined by Captain Howson in October 1936: in addition to the routine office work of the Society and the three breed societies and the preparations for the show itself, it fell to them to compile the annual *Journal* and, normally, a stud and herd book, and to administer the War Office premiums and Racecourse Betting Control Board grants in respect of Welsh cob and pony stallions, to arrange tours of inspection of stallions, and to collect and tabulate service and foaling returns. Howson had personally attended forty-seven meetings of Council, committees and sub-committees during the year, which entailed the drafting, stencilling, collating and dispatching of many reports and sets of minutes in connection with meetings of the breed societies. In addition, he calculated that between 30 May and 24 June 1936 the work associated with the Abergele show, which was done on Saturdays, Sundays and after 6 p.m. in the evenings, totalled at least nine weeks, on the basis of eight hours' work per day.

While the Society agreed to the request for a permanent new appointment, a long discussion ensued at the Finance Committee meeting of November 1936 as to whether the person recommended by the secretary and the honorary director, but who had no knowledge of Welsh, should be

12. Reuben Haigh of Ruabon, honorary director from 1930 to 1950.

appointed. Finally the meeting approved the motion moved by Moses Griffith and seconded by T. J. Jones that the job advertisement should stipulate that a knowledge of Welsh was a desirable qualification. Accordingly, a little later the Society appointed 24-year-old Eryl Glyn Roberts, a Welsh-speaker from Chester to the post.

By its very nature as a national society the Royal Welsh Agricultural Society was required to come to terms with the delicate matter of the relative status accorded to Welsh and English in its various activities. While English was from the beginning the dominant language of the Society's administration and its annual show, Welsh articles were printed and editorial notes were bilingual in the successive numbers of its *Journal* from 1904 onwards. Moreover, at the Council meeting of February 1912 Thomas Whitfield of Shrewsbury, on being appointed to the vacant post of secretary, undertook to learn Welsh and to engage a Welsh-speaking correspondence and committee clerk as well as a Welsh-speaking surveying and quantities clerk. Whitfield did not, however, remain long in his post, and a new secretary was sought in autumn 1915. And although David Davies, as chairman from 1921, wrote his annual *Report on the Show and the General Working of the Society* in English down to 1929, from 1930 this was presented in both languages. In the period after the First World War, when Welsh patriots knew their language to be in decline and under threat, it was only natural that some members of the Society would call for a greater use of Welsh in its affairs. In December 1931 Kenneth Davies of Cardiff suggested that greater prominence might advantageously be given to the Welsh language at the Society's shows by printing the catalogue, and having the signposts painted, in Welsh as well as English, but nothing came of this. Moses Griffith, as we have shown, was to the fore in securing a Welsh-speaking member of the office staff from 1937 and after the war he was to make a more robust stand for the use of Welsh in the running of the Society's affairs.

In these inter-war years the Royal Welsh Agricultural Society, largely for financial considerations, became increasingly hostile to the periodic practice of the Bath and West of England and Southern Counties Society holding its shows in Wales and Monmouthshire. Upon receipt of information in December 1924 that the Bath and West had approached the Cardiff City corporation seeking an invitation to hold its show in the city, the Society's Council instructed the chairman to inform Cardiff's lord mayor of its opposition to the visit. Whether this had any impact or not, the happy outcome was that Cardiff decided not to entertain the Bath and West in 1927 given that the Royal was visiting Newport in the same year. In November 1931 the Society was again informed by its chairman that a movement was afoot to persuade Swansea Town Council to issue an invitation to the Bath and

West to hold its show there in 1933; as he had done previously, David Davies went on to point out that the holding in Wales of any leading English show, other than the Royal, would be injurious to the interests of the Royal Welsh Agricultural Society and he urged that every effort be made to resist the incursion of the Bath and West into an area 'which could only rightly be regarded as the province of the National Society of Wales'. There followed resolutions of protest by all county meetings of the Royal Welsh and by the Council of the Welsh Black Cattle Society against the Bath and West visiting the territory of the Royal Welsh in future and the opposing points of view of both the Royal Welsh and the Bath and West were aired in the press. The campaign paid off, for Swansea Town Council decided against inviting the Bath and West to the town in 1933.

A losing battle was fought by the Society, however, in its effort to prevent the Bath and West Show visiting Neath in 1936. Having registered its protest and argued its case before members and the wider public on the grounds that Wales was not large enough to accommodate two major shows in one year – Lord Davies referred in July 1934 to poaching by 'foreign societies' – it was decided in mid-October 1934 that no further action be taken. For its part, Plaid Genedlaethol Cymru, the National Party of Wales, decided to make a protest at the Neath show in June 1936 and this caused some embarrassment to the Society, anxious as it was throughout to avoid any involvement with political parties.

No such protest was directed against the visit of the Royal Society Show to Wales, on a formal level at least. Yet influential voices within the Society itself and in wider Welsh agricultural circles pointed to the harm done to the Royal Welsh by such incursion into the large industrial towns of south Wales. Speaking at the Swansea show in 1927, the earl of Dunraven reminded his listeners that the event had been affected by a very wet summer and by serious competition from the Royal Show at Newport in early July. Some criticism, too, was to be heard over the decision of the Royal Welsh Agricultural Society to suspend its show on the occasion of the Royal visiting Cardiff in 1938, notwithstanding its realization that a financial loss would be incurred. No less a respected person than W. H. Woodcock, show reporter for the *Western Mail*, mused in print a year later: 'One wonders whether it is advisable for these English and Welsh institutions to combine forces at any time, for some argue that it has an adverse effect upon subsequent financial support and general enthusiasm for the Welsh Show.' That the official position, however, was one of maintaining cordial relations with the Royal Society was to be seen in 1936 when the Council decided to ignore the protest by Miss Mallt Williams of Pant-y-Saeson, St Dogmaels, at its decision to withhold the show in 1938.

For all the problems of a sluggish membership, itself a reflection of public apathy, and a slender reserve fund of less than £10,000, the Society nevertheless made advances in the years between 1910 and 1939, especially from 1922 onwards. Its organization was strengthened and democratized, above all through the development of county committees. Moreover, a measure of financial security was guaranteed with the commencement of a reserve fund, although this had to be raided at the close of the 1930s to cover the deficits of the shows of 1936 and 1937. A worry for the Society's officials was that everything in these years rested upon the revenue generated by the shows, and it is to the show that we must now turn our attention.

CHAPTER FIVE

'As Good a Show as there is in the Kingdom'

In spite of the underlying weakness of the Society's financial position, such was the growth of the show between 1910 and 1939 that its standing came to equal and in some respects surpass that of the other large exhibitions within the United Kingdom. The *Western Mail* commentator, Charles E. Lloyd, enthusing on the Swansea show of 1927, reflected: 'I am of opinion that the Welsh National is as good a show as there is in the kingdom. I say this after making a round of all leading shows.' Chairman David Davies was likewise to claim in his *Report* for 1927 that the Royal Welsh Show was able to take its rightful place amongst the most important in Britain and that it was universally acknowledged as one of the four national meetings of the kingdom. Certainly the increasing support it enjoyed during the 1920s and 1930s not only from within Wales itself and the immediate border counties but from the leading agriculturalists, stockbreeders and traders of the United Kingdom as a whole testified to its enhanced status and recognition as one of the most important events of its kind in the British Isles. At Carmarthen in 1925 entries came from all parts of the United Kingdom – from Sussex and Essex in the south-east to the Scottish Highlands in the north – while ten years later entries from such distant counties as Lancashire, Staffordshire, Surrey, Dorset and Cornwall were on show at Haverfordwest.

Local pride and competitiveness willing its *own* to be the biggest and best yet ensured that shows grew larger over time. The compelling urge to outdo was not simply a matter of south versus north, as, for example, was the case in 1924 with south Walians hoping that their show at Bridgend would take away the gate record set at Wrexham two years earlier; it prevailed at an inter-regional level, with the local officials of the forthcoming show at Abergele in 1936, for instance, making it clear to all and sundry that the entries for Llandudno in 1934 were so far exceeded as to make comparison odious!

No significant increases, however, on the exhibitions at Aberystwyth occurred in the scale of the migratory shows held between 1910 and 1914 if we compare the total number of livestock entries. Whereas these averaged 563 over the six shows at Aberystwyth, livestock entries at Llanelli, Welshpool, Swansea, Porthmadog and Newport over the years 1910–14 averaged only 597. On the other hand, attendance rose significantly, although we need to bear in mind that the figures were often estimates. Peaking in 1908 at Aberystwyth with 15,000 visitors, Llanelli's attendance in 1910 was estimated at 25,000, while 35,000 were said to have visited Porthmadog in 1913.

Taken in the round, however, the annual migratory show was to become a much larger and increasingly prestigious event only after its resuscitation at Wrexham in 1922. That show, under the skilled organization of Arthur Evans and which continued as the three-day event it had already become at Newport in 1914, was a magnificent success when compared with the pre-war shows. Nothing approaching the 1,140 livestock entries at Wrexham had ever been seen on a Welsh National Agricultural Society showground hitherto, and their uniformly high quality noticeably surpassed the standard of livestock at any previous Welsh National Show, which was a testimony to the great advancement in breeding which had been achieved by Welsh farmers during the war years. Traders' stands, or 'spaces' as they were referred to, were generally regarded from the beginning as a reliable criterion of the extension or decline of the show: totalling 127, the number of traders' stands at Wrexham easily outstripped those of the pre-war years, which had numbered 84, 64 and 78 at Swansea, Porthmadog and Newport respectively in the immediate pre-war years.

Reference to both Table 2 and Appendix 1 will reveal that the altogether grander exhibition staged at Wrexham was to be maintained and sometimes surpassed (though not in terms of total entries) in the years that followed down to 1939. Attendance over these years averaged approximately 35,000. No show, apart from the staggering gate at Carmarthen in 1925 of nearly 53,000, exceeded the 40,000 mark. At the other end of the scale, the disappointing attendance at Monmouth in 1937 of 22,575 was largely the result of inclement weather, a factor which played a big part in depressing gates at Swansea (a one-off, four-day event) in 1927, Llanelli in 1931, Abergele in 1936 and Caernarfon in 1939. The number of traders' stands, too, was maintained following the new standard set in 1922. There is no mistaking, however, the difficulties facing commerce from the close of the 1920s into the early 1930s and, accordingly, with the return of a better business climate from mid-1933 there occurred a significant rise in the number of trade exhibits at the shows held between 1934 and 1937. Similarly, entries in the livestock and other competitions remained fairly buoyant after 1922. In a very real sense the Royal Welsh Show was, as it has been since, a livestock exhibition and entries in the various livestock classes and sections were, after rising appreciably at Carmarthen in 1925, to peak again at the three shows at Llandudno, Haverfordwest and Abergele in the mid-1930s (see Appendix 1). Each of the shows between 1910 and 1939, of course, had its own distinct characteristics and achievements, but some were singled out by contemporaries for special

Table 2. Comparative statement of entries, traders' stands and attendance 1922–1939

YEAR	SHOW	LIVESTOCK ENTRIES	TOTAL ENTRIES (INCL. LIVESTOCK AND TRADERS' STANDS)	TRADERS' STANDS	ATTENDANCE
1922	Wrexham	1,140	4,189	127	(est.) 30,000
1923	Welshpool	1,103	3,123	118	30,032
1924	Bridgend	1,111	4,105	130	28,316
1925	Carmarthen	1,343	3,321	128	52,731
1926	Bangor	1,050	3,360	137	38,573
1927	Swansea	972	3,777	115	38,387
1928	Wrexham	1,271	3,490	164	29,867
1929	Cardiff	1,077	2,524★	172	36,917
1930	Caernarfon	1,101	2,651★	124	37,506
1931	Llanelli	967	2,588★	141	39,930
1932	Llandrindod	1,053	2,179	139	26,519
1933	Aberystwyth	1,166	2,829	131	39,837
1934	Llandudno	1,323	3,226	160	39,037
1935	Haverfordwest	1,422	3,480	154	38,847
1936	Abergele	1,501	3,620	161	34,105
1937	Monmouth	1,192	2,783	161	22,575
1939	Caernarfon	1,079	2,610	142	30,068

★ No dog show

praise: Carmarthen, despite its unfavourable weather, was commended by the *Western Mail* for its record-breaking figures, a fitting celebration, indeed, of the Society's coming of age; the Llandudno show of 1934 was described by the same newspaper as the finest since the First World War; while Haverfordwest's exhibition a year later was acclaimed as the best for a decade.

When its annual show became migratory the Society faced a fresh set of challenges. Foremost, of course, was the choice of future venues. Upon receiving deputations from a small number of towns inviting it to hold the show at their centres, the Council normally reached a decision after weighing the relative attractions of the terms of the competing bids. Various factors such as the general suitability of the area, the size of the proposed site, its railway accessibility and water supply, and the willingness of the applicants to comply with the Society's financial requirements were considered. With regard to the latter factor, it was a rule that the local committee formed to handle the show arrangements should draw up a local agreement that would either pledge a guarantee of £1,000 or, if it preferred, donate £500 (raised to £1,000 after 1927) to the funds of the Society. Furthermore, a proportion of any balance – stipulated as 75 per cent in early 1927 – arising from the local accounts after all charges in connection with the show had been settled was to be paid into the Society's show account. Sometimes difficulties arose over both these requirements. For example, in the aftermath of the 1928 show at Wrexham the Society had to instruct its secretary to communicate with the town clerk of Wrexham with a view to securing payment of the guarantee and that, if a satisfactory outcome was not achieved, he should take legal advice regarding the Society's position in the matter. Less than two months later, the Council was informed by the Cardiff local committee, which was responsible for the forthcoming show in the city in July, that in so far as the disposal of any balance remaining in the local fund was concerned it should be retained by the local committee and utilized to any purpose they might think fit. In support of this view they drew attention to the fact that this was the customary procedure in the case of the Royal, the Bath and West and the Southern Counties shows. Defending his Society's position, the chairman in reply urged that the Royal Welsh Agricultural Society was a youthful organization which did not enjoy the security of substantial reserve funds and that if the show was to be established on a permanent basis it was essential that the Society should amass a reserve fund adequate to meet any contingency which might arise. As remarked in the previous chapter, it was precisely this want of reserves which constrained the Society from reducing the size of the donation or guarantee required from localities visited by the Royal Welsh Show, thereby preventing smaller towns from issuing an invitation. Corwen had made a bid to host the 1932 show, but its inability to comply with the Society's financial requirements led to rejection out of concern on the part of the chairman that an unfortunate precedent might be created. In response to the increase of the reserve fund to more than £10,000, calls for a reduction in the donation or guarantee required were made by some county meetings in 1934 but rejected on the grounds that the time was not yet opportune.

Vitally important in ensuring a successful visit of the Royal Welsh Show to any town or district was the readiness of local agricultural societies in those areas to withhold their own show. In spite of the fact that certain privileges were granted to members of these local societies in connection with the Royal Welsh Show such as free admission and reduced entry fees, withholding their own shows entailed financial loss; it was sometimes no easy matter to revive a show which had been allowed to lapse for even one year, and a local society invariably experienced difficulty in collecting subscriptions for any year in which it did not hold a show. Given the very real losses sustained by the local societies, it is hardly surprising that some contention arose in the wake of the Abergele show in 1936, for the Denbighshire and Flintshire Agricultural Society had in fact withheld its show on two occasions within a very short time to facilitate the Llandudno show of 1934 as well as the Abergele one.

The whole subject of the best policy to pursue with regard to the location of the show was discussed at a Council meeting in late July 1927 at the

Swansea showground when C. Bryner Jones raised the question of venues for the 1930 and subsequent shows. Intimating that his views were shared by others on the Council, he expressed his doubts whether large industrial centres were the best places for holding future agricultural shows and he wondered whether the time had not come for the Royal Welsh to consider holding some shows in the more purely agricultural centres of Wales. Holding up Corwen as a possible venue for the show by counting in its favour its railway facilities and the fact that a successful National Eisteddfod had been held there, he suggested that visiting the more purely agricultural areas might awaken a wider interest in the Society and lead to an increase in membership. As we have noticed earlier, its cautious chairman, David Davies, while sympathizing with the principle of holding the show in the more rural areas, believed that, given the Society's slender financial reserves, it might prove to be too hazardous a venture to undertake. Although the Society would set its face against reducing the size of the donation or guarantee required from centres visited by its show, from 1930 onwards it started to take risks in visiting certain areas. The agricultural columnist W. H. Woodcock of the *Western Mail* thus commended the Society in July 1932 on its 'pioneering spirit' in choosing to go to Llandrindod Wells that summer. In view of its comparatively inaccessible situation and the sparse population of the area, a willingness to open up new ground was vindicated, he maintained, in the second-day gate of just over 16,300. Perhaps he was being over-sanguine, for the total attendance over the three days of 26,519 was low compared with past and future shows. Lord Davies explained that in accepting the invitation to Llandrindod Wells the Council felt that, although the likelihood of making a big profit was remote, it was the Society's duty to visit a sparsely populated rural area which had not previously hosted the show. Confronted with the heavy financial loss on the show, he put a brave face on it by claiming that it was widely held that the missionary and propaganda value of holding the show in Radnorshire could not be overestimated. The Society was equally apprehensive about the financial prospects of the Haverfordwest show held in 1935, fearing that the remoteness of the show venue from the English border would deprive the exhibition of many distant competitors and visitors who had been important to the Royal Welsh's success in the past. As things turned out, the anxiety proved to be groundless, although the low entries from north Wales in the Welsh cattle section may well have been the consequence of the long journey involved.

This last factor of distance between north and south was highlighted by the *Western Mail* in 1936 as a serious drawback. Commenting on the very poor representation of south Wales in the show rings and pens at Abergele that summer, it observed: 'There is just one disadvantage with a show that is held in North and South alternately, and that is that the cost of transportation and the expenses incidental to exhibition are prohibitive for breeders in the area furthest away from that in which the show is being held.' Other shows certainly experienced this defect: at Porthmadog in 1913 there were practically no entries from south Wales in the cattle classes and few in the sheep; the Carmarthen show in 1925 saw disappointing entries in the County Group Competitions for native Welsh breeds which were strongest on the ground in the northern counties; and the south was again poorly represented in the cattle classes at the Caernarfon show in 1930. To a degree at least, the earlier forebodings on the part of some like Loveden Pryse that the separation between north and south would come about if the show left Aberystwyth were being realized.

The peculiar difficulty and expense of travel in Wales meant that, in these years when motor transport was still limited, cheap rail facilities in connection with the show were an important consideration for the Society, even more so than for other of the big agricultural societies. It meant frequent haggling over passenger fares and charges for carriage of livestock with the railway companies. Concern among members at this unfair treatment was reflected in the universal request of the county meetings in 1928 that the Council should again approach the railway companies, strongly urging that the same facilities be accorded to members of the Royal Welsh Agricultural Society attending their show as had been accorded to those of the Royal Agricultural Society of England and of the Smithfield Club when visiting their shows. Renewed representation proved fruitful, and further efforts by the secretary also succeeded in getting the cheap travel facilities enjoyed by members extended to exhibitors at the 1929 Cardiff show. However, dissatisfied that a mere 121 persons had availed themselves of this concession, the railway company withdrew the cheap travel facilities for members and exhibitors attending the Caernarfon show in 1930. Persistence on the Society's part, nevertheless, gained period travel facilities at reduced rates for members and exhibitors visiting the Llanelli show in 1931, on the understanding that the Society should make every effort to ensure as large a number of visitors as possible availed themselves of the concession offered. Moreover, the railway company on this occasion also agreed to reductions in what were viewed by members as extortionate charges made by the railway companies for cartage of stock from the railhead to the showyard.

As had been the case during the Aberystwyth years, the precise date of the show remained an important consideration for the officials. Down to the close of the 1920s, with the exception of those at Porthmadog (1913) and Newport (1914), the shows had been held in August Bank Holiday week on the grounds that it assured a good 'gate'. From 1930, however, the show would no longer

13. The prince of Wales at the Welshpool show, 1923.

be held in early August, the Council being aware of the difficulty in obtaining visits from members of the royal family at shows held after the end of July and also wishing to steer clear of holding the show during the same week as the National Eisteddfod. Avoiding a clash with other important agricultural shows also became a major consideration in the precise dating of the show from the early 1930s. As early as 1927 Captain Howson had referred to the ill temper manifested within the United Kingdom as a whole over the clashing of show dates. It is no surprise, therefore, that from 1932 onwards Council was to be swayed in fixing the dates of forthcoming shows by its wish to avoid clashing with both the Yorkshire Agricultural Society Show, which was held in early July, and that of the Royal Lancashire Agricultural Society, which fell at the beginning of August. 'Available' to the Royal Welsh Agricultural Society for holding its show were therefore the third and fourth weeks of July, and Council invariably opted for the final week.

The desire to set a date that would enable a royal visit was recognition of the tremendous drawing power which the royal presence exerted on the showground. Patronage was continued after King George V's death in 1936 by his successor King Edward VIII, who, as prince of Wales, had earlier accepted the presidency in 1922 and 1923, thereafter acting as honorary president. As well as being a frequent exhibitor at the Royal Welsh, Edward had attended the shows at Welshpool and Llanelli in 1923 and 1931 respectively. The duke of Kent visited the show at Aberystwyth in 1933, and in 1936 intimated his acceptance of the office of honorary president in succession to the duke of Windsor. The Carmarthen show of 1925, too, had been rendered additionally attractive by the visit of Prince Henry.

Of course the best-laid plans in terms of show dates and the choice of a suitable venue – bringing with it in many instances a naturally beautiful site like Penrhyn Park, Bangor, in 1926 and Singleton Park, Swansea, the following year – could be foiled by bad weather and other unforseeable factors. Truly appalling were the second and third days of the Caernarfon show in 1939, with many lifelong showgoers declaring that the last day of the event, with its mist, high wind and a steady downpour, was one of the worst they had ever experienced. Foot and mouth disease, too, adversely affected several shows in the period 1910–39. Swansea's 1912 show was faced with both foot and mouth disease orders and swine fever, which meant that, although the cattle

14. The handsome entrance, built in the form of a medieval castle gate, to Penrhyn Park, Bangor, home of the 1926 show.

classes survived practically intact, entries in those of sheep were drastically reduced and there were no pigs in the showyard. By some perverse quirk of fate Swansea's show in 1927 likewise experienced a small drop in entries because of the presence of foot and mouth disease in parts of the United Kingdom, the cancellation of the entry of a large number of cattle from the afflicted Tunbridge Wells district serving as one instance of its unfortunate effect on the show. On this occasion, inclement weather and the depressed state of trade also conspired together to harm the event. On the eve of the Wrexham show the following year the foot and mouth scourge appeared in Lancashire, though, happily, only the Friesian classes at the show were adversely affected. Although the Llanelli show in 1931 was spoiled by the heavy rainfall of the second and third days, the threatened abandonment of the cloven-hoofed sections following a widespread outbreak of foot and mouth disease shortly before the show was fortunately averted by the lifting just in time of the countrywide embargo on the movement of livestock. So persistent were the outbreaks of foot and mouth disease at the height of the different show seasons that Captain Howson ventured to query in 1928 whether such outbreaks really were spontaneous and matters of pure coincidence, or whether they were not caused by movements of livestock around the country. With visits by a member of the royal family helping to draw in the crowds, there is no mistaking the despondency among showgoers at Caernarfon in 1930 at the unavoidable failure of the prince of Wales to attend the show. Bad weather conditions had prevented the prince from flying his own aeroplane to Caernarfon for the opening day, and assurances given that he would visit the showground on the following Friday were again undone by the weather, his plane having to turn back at Birmingham in the face of dangerous flying conditions. Disappointment was felt, too, when the duke of Kent's visit to Abergele in 1936 was cancelled because of unavoidable circumstances.

First among the Royal Welsh Agricultural Society's aims was to improve the quality of the native livestock – Welsh cobs and ponies, Welsh Black cattle, Welsh Mountain and Kerry Hill sheep, and Welsh pigs – and an important new step towards achieving this was taken in 1924 as we have earlier intimated. Not surprisingly the Inter-County Breed Competitions instituted as an annual show event from that year onwards were the brainchild of the chairman, David Davies, a keen supporter of the native breeds. Their introduction was of huge significance for they constituted the one feature of the Royal Welsh Show which was unique; it represented a departure from ordinary show routine within the wider kingdom as a whole. These competitions were only made possible by the fact that Wales alone within the British Isles possessed its own distinctive national breeds of all the major kinds of livestock – horses, cattle, sheep and pigs. The 'County Competitions', by fostering so much keen rivalry, served both to improve the quality of the native breeds and to advertise their merits. Until 1930, competitions between the counties were confined to the native breeds, with each Welsh county and Monmouthshire sending representative groups of native breeds in six groups – Welsh Black cattle, Welsh cobs, Welsh ponies, Welsh pigs, Welsh Mountain sheep and Kerry Hill sheep – and collecting points in each division. The prize, which occasioned so much jealous pride in the counties, went to the one with the most points on its aggregate entry. Reflecting the importance of the feelings of the county meetings in shaping developments, the Society responded to wishes expressed on their part in 1929 by reorganizing the County Group Competitions at the 1930 show. The desire of five county meetings for an amendment of the rules governing the County Group Competitions sprang from a recognition of the lack of competition under the existing rules since most breeds were debarred from competing. The five called for a widening of the scope of the competitions in order to enable every county to show a representative group. Carmarthenshire went so far as to complain that, as the competitions were at present constituted, most breeds were debarred from competing and that it was felt that they were alienating sympathy and doing more harm than good. Seeking to preserve the native

15. Welsh Black Champion Group, Denbighshire Group, 1927.

breed competitions intact, the Society's solution was to split the County Group Competitions into two distinct sections. The first was devoted solely to Welsh breeds and the competition continued on exactly the same lines as hitherto. In the second section, the prize was awarded to the county gaining the highest aggregate points in the ordinary open classes for horses, cattle, sheep and pigs, irrespective of the breed of the successful animals.

As was noted earlier, it is significant that when faced with pressure to economize on the show's schedule during the lean years of the early 1930s, the Society decided to retain Scheme A for Welsh national breeds at the successive shows from 1932 to 1935, and to exclude Scheme B for all British breeds within Wales. As a distinctive feature of the Royal Welsh Show, every effort was made to safeguard Scheme A; David Davies volunteered to give £50 out of his own pocket towards the prize money required for its inclusion in the 1932 Llandrindod Wells show schedule if the remaining £50 could be raised elsewhere. Particularly keen, too, in promoting the improvement of the national breeds was D. D. Williams, who played a vital part in organizing the competition of the County Group of the native breeds of Wales at successive shows. Important as Scheme A undoubtedly was in encouraging improvements to the native breeds, it was none the less the case that inter-county rivalry relating to the purely national classes, particularly the Welsh Black cattle, was most keenly felt in the north, the geographical stronghold of the native breeds. It followed that this intense rivalry was most in evidence in the shows held in the north. This being so, it is not surprising that four of the five county meetings calling in 1929 for a change to the rules governing the County Group Competitions were located in the south.

Writing in the *Western Mail* in July 1931 W. H. Woodcock observed that the Royal Welsh Agricultural Society had adopted a very wise policy in providing district classes, in addition to open ones, for various breeds of livestock at its annual shows and later, in 1937, he was to comment that both the district classes and the Welsh Inter-County Group competition were 'distinctive features' of the Royal Welsh Shows. District classes were confined to exhibitors residing within the particular county and neighbouring ones wherein the show was located and were meant to compensate local agriculturalists for withholding their own agricultural societies' shows when the Royal Welsh visited their district. Such local or district classes were in operation from the early days of the migratory show and during the inter-war years, notwithstanding the disappointing level of support between 1922 and 1926, they were to become a major feature of the second day of the show. Although they lacked the dignity and quality of the aristocratic beasts of the first day's open classes, these district classes of livestock were of greater interest to the local farmers and visitors since there was a sense of local rivalry in the competitions.

Moreover, they gave the 'novices' a real opportunity of winning the coveted prize cards as well as the prize money, which would have been far less likely in open competition among the experts of the showing world.

The open classes attracted exhibits from breeders throughout the kingdom, and new breeds made their appearance on the showground at different times during the inter-war years: the cattle sections thus saw the introduction of British Friesian cattle in 1922, Aberdeen Angus cattle in 1924 and 1934, and Guernsey cattle in 1935; similarly, Suffolk sheep were introduced in 1924 and both Southdown and Wiltshire sheep in 1926. Goats appeared for the first time in a Royal Welsh Show schedule in 1928. Even so, year after year the leading features of the show comprised Welsh Black cattle, Herefords and Shorthorns – the latter two English breeds popular in Wales especially in the mid-south and western districts – Welsh Mountain sheep and Kerry Hill, native Welsh pigs and Welsh ponies and cobs. While entries of in-hand cobs at the shows of the 1920s – fluctuating between twenty and thirty – were lower than in earlier years because of the decision to debar hackneys, the plunge in entries during the mid-1930s was a cause for concern. Doubtless this falling away was a reflection of the decline in numbers of good Welsh cobs brought about by the growth of other means of locomotion and to some extent, also, by the reduction during this decade in premium grants for Welsh cobs and Mountain ponies from the War Office and the Treasury as part of their economy campaign. From the disappointing turnout of Welsh cobs in 1934 and 1935, matters took a decided turn for the worse in 1936 when there were only ten of them all told! Following a gradual curtailment of the classes for their breed at recent shows, classes at Monmouth in 1937 were reduced to just two and, once again, the entry was extremely disappointing, for only three mares, one of which failed to turn up, and five stallions, one, again, an absentee, were catalogued. Poor levels of cob entries from the mid-1930s were paralleled by those in Welsh Mountain ponies and Welsh ponies, with just thirty-one brought forward in the two sections in 1937, and two of the Welsh pony classes cancelled owing to insufficient entries.

There is no mistaking the steady improvement in the quality of livestock exhibits taken overall at the Royal Welsh Shows over the period 1910–39. Nor should this surprise us, given that the standard of breeding within the British Isles was improving in the 1920s and 1930s under the joint impetus of the breed societies, the livestock improvement scheme of the Ministry of Agriculture, the ready market for highly bred stock at home and abroad, and the annual agricultural shows themselves. The Welsh pig, in particular, was to make big strides during the inter-war years towards becoming established as one of the best types of bacon pigs in the United Kingdom, in the process benefiting greatly from the encouragement of David Davies of Llandinam. Bringing all the winners – proud lords of the ring – together at the Royal Welsh Shows was the Grand Parade, which in the view of Charles E. Lloyd of the *Western Mail*, writing in 1926, was rarely equalled in any other show.

Certain exhibitors and their superlative livestock specimens stand out in these years, some of them, indeed, attaining legendary status as true aristocrats. Wern Sentry, the Welsh Black bull owned by R. M. Greaves of Porthmadog, was champion at the Royal Welsh as well as at all the leading shows for several years, taking the championship on no fewer than five successive years at the Royal Welsh between 1924 and 1928. Impressive, too, in

16. Welsh Black bull, Wern Sentry, winner of the Colonel Harry Platt Memorial Challenge Cup at the 1928 show.

17. The Welsh Black bull, Egryn Buddugol, at the 1932 Royal Welsh Show.

winning this same Colonel Harry Platt Memorial Challenge Cup was Moses Griffith's Egryn Buddugol, a champion in 1932, 1933, 1934 and 1935. After being beaten in 1936 by an Anglesey bull, he made a triumphant comeback in 1937 at nine years old. Other prominent winners in the Welsh Black classes were Lord Penrhyn, David Davies of Llandinam, Captain Bennet Evans of Bow Street, The Hon. Lady Shelley-Rolls of Monmouth, Mrs Williams-

Owen of Treveilyr, Bodorgan, Anglesey, and, as bona fide farmers, J. M. Jenkins of Tal-y-bont, D. W. Morris of Tal-y-bont, David Jenkins of Taliesin, Cardiganshire, and Richard Rees of Pennal, Machynlleth. A real triumph was achieved by J. M. Jenkins at the Abergele show in 1936 in winning the Mathias Challenge Cup, the male championship and the supreme breed championship with his grand yearling bull Caran Penda, the female championship and reserve for the breed championship with his heifer Caran Jano and the Edward Prince of Wales Challenge Cup for the best group. Noteworthy, too, was D. W. Morris's Penywern Hester, who took the female championship in 1933, 1934 and 1935. Widely known as a successful breeder of Welsh Black cattle and Welsh Mountain sheep, Morris's young death at the age of forty-four in 1937 was a blow to Welsh farming.

By the 1930s Shorthorn were the most popular breed reared in Wales. Prominent exhibitors were the blind Lord Merthyr of Hean Castle, Saundersfoot, Pembrokeshire, before his death in 1932; G. E. FitzHugh of Plas Power, Wrexham; Richard Stratton of The Duffryn, Newport; Colonel Sir Edward Curre of Itton Court, Chepstow; Daniel Beynon of Ynyshafren Farm, Ponthenry, Llanelli, whose bulls were outstandingly successful at the shows between 1929 and 1932; Earl Cawdor of Stackpole, Pembrokeshire; Captain N. Milne Harrop of Ruthin; Llysfasi Farm Institute, Ruthin; Major G. Miller Mundy of Andover; G. H. Willis of Birdlip, Gloucestershire; Lieutenant Colonel E. C. Atkins of Hinckley, Leicestershire; E. Uwins Gillate of Surrey; and Miss R. M. Harrison of Staffordshire, whose Dairy Shorthorn bull, Townend Supreme, was notably successful at the 1936 and 1937 shows. Winning Shorthorns were also sent by the king, the prince of Wales from his Cornish herd, and by the duke of Westminster, Eaton Hall, Chester.

A successful exhibitor in the Hereford section in the 1920s was D. P. Barnett of Llancarfan, Glamorgan, whose great Royal champion Apsam was an easy winner in the first bull class at the 1924 show. Other successful Welsh exhibitors of Hereford cattle at the shows in these years were Sir David Llewellyn of St Fagan's with bulls like the massive and imposing St Fagan's Paxolute in 1928 and St Fagan's Pandarus in 1930, and, in the 1930s, James Price and Son, Glantowy, Llandovery, J. L. M. Sinnett of Tal-y-bont-on-Usk, Breconshire, T. E. Gwillim of Talgarth, D. G. P. Jeffreys of Trecastle, Breconshire, and, from further west in Pembrokeshire, Alan Colley of Corston, and T. H. Scurlock of Tiers Cross, Haverfordwest. Of course, many of the big Hereford exhibitors were from the border counties, notably Percy Bradstock of Tarrington, Herefordshire, W. H. Brown Cave of Leominster, Henry Dent of Perton Court, Hereford, John Parr of Ross on Wye, F. J. Newman of Leominster, Craig Tanner of Wroxeter and H. R. Griffiths of Little

Tarrington, Herefordshire. Outstanding among Griffiths's animals was the magnificent Britannia, breed champion at the Royal Welsh in 1933 and 1934, as well as champion female at the Royal in 1933 and 1934, and, according to the knowledgeable Captain Howson, 'undoubtedly the greatest show cow of her time'. Another superlative specimen to the point of being an almost perfect example of his kind was Percy Bradstock's Free Town Admiral, who, when winning Hereford old bull in 1932 at the Royal Welsh, had won the Royal championship that year for the third time. High-quality Hereford entries, too, came from the king's world-famous Windsor herd, like the weighty, short-legged bull Sultan, which won the senior bull at Monmouth in 1937.

18. Percy Bradstock's Free Town Admiral, winning Hereford old bull at 1932 Royal Welsh show.

Welsh Mountain sheep understandably attracted most entries in the sheep section. Prominent exhibitors in the 1920s and 1930s were David Price of Nantyrharn, Cray, Breconshire, Major Eric Platt, Madryn Farm, Aber, Caernarfonshire, University College of North Wales College Farm, Llysfasi Farm Institute, the aforementioned D. W. Morris of Tal-y-bont, Lieutenant Colonel E. W. Griffith, Plasnewydd, Trefnant, Johnny Morris of Senny-bridge, David Lloyd of Tremeirchion, St Asaph, and G. J. Thomas of Carregcegin, Llandeilo. David Price won the breed championship in 1931 and 1932; his winning ram in 1932, Nantyrharn B3 3596, was of exceptional merit. Outstanding rams, too, were Snowdon D57, winner of the supreme championship in 1934, Snowdon G5, a Royal winner, who headed the old rams at Abergele in 1936 and took the supreme championship in 1937, both of them owned by the University College of North Wales, and Cegin M14, belonging to G. J. Thomas of Llandeilo, which took the breed championship in 1939. Thomas also won the Black Welsh Mountain sheep breed championship at the same show with his Cegin Wonder. Other notable exhibitors and winners of the Black Welsh Mountain classes during these years were Mrs B. A. Jervoise of Herriard Park, Basingstoke, Major General Lord Treowen of Llanover, Abergavenny, Ewart Owen of Prestatyn (who was also a notable winner of Welsh pig classes) and Major J. A. Herbert of Llanover. Another principal native breed, the Kerry Hill, saw its classes dominated in the late 1920s and early 1930s by John Beavan of Winsbury, Chirbury, but the 'Winsbury Wizard's' erstwhile supremacy was to be successfully challenged in the ram classes from 1934, in particular by the very masculine and massive Kerry Goalkeeper belonging to Thomas Williams of Forden, Welshpool, this great show ram winning reserve champion in 1934 and the breed championship at the next two shows. Prominent and successful exhibitors, too, in the late 1930s were J. W. Owens of Shobdon, Leominster, and the duke of Westminster.

19. The champion Kerry Hill ram, Kerry Goalkeeper.

Another star, this time in the horses section, was Cwmcau Lady Jet, a brood mare in the Welsh cob department belonging to Messrs John Jones and Son of Dinarth Hall, Colwyn Bay, which won first prize at Llandrindod Wells in 1932, Aberystwyth in 1933 and Llandudno in 1934, and runner-up at Haverfordwest in 1935. Likewise, the Welsh Mountain pony stallion, Grove Sprightly, owned by the celebrated breeder Tom Jones Evans of Lower Dinchope, Craven Arms, won his class at all the Royal Welsh Shows – apart from the 1937 one when his owner was the judge – stretching from 1930 to 1939! Grove Sprightly also won all his harness classes at the Royal Welsh from 1931 to 1935 inclusive. Such was his unbroken run of success and his appeal to ringside crowds that he was an 'Idol of Idols' in the opinion of seasoned showgoer William Evans writing in 1961, who indeed saw him as equalled in another section and at a later date only by Pentre Eiddwen Comet. 'A real aristocrat' was the description given Montgomeryshire breeder H. Meyrick Jones's Mathrafal Eiddwen by a commentator in the *Western Mail* upon his winning the George, Prince of Wales Challenge Cup for the best cob of the old Welsh type at the 1929 show, a silver cup he had also won earlier in 1926 and 1927 and did so again in the following year, 1930.

20. Welsh Mountain pony stallion, Grove Sprightly, owned by Tom Jones Evans.

21. Welsh Cob stallion, Mathrafal Eiddwen, at the 1927 Swansea show.

Throughout its history the Royal Welsh Show has benefited from the support and patronage of both the royal family and well-to-do landed families. From their ranks were drawn many of the shows' exhibitors and prize-winners, as has been demonstrated in the foregoing pages. But smaller farmers, too, including tenant farmers, were sometimes successful exhibitors in the open classes. At the 1923 show at Welshpool it was notable that in both Welsh Mountain sheep and Welsh Black cattle classes, many of the successful exhibitors were comparatively small tenant farmers. Even more so, the opening day at the 1930 show at Caernarfon essentially belonged to the farmers, their victories pleasing both farmers and breeders alike, but none more so than David Lloyd George, who stayed on the ground for two hours, and characteristically enjoyed the rout of the big exhibitors by the local farmers. The premium bull scheme whereby animals selected by government livestock officers were bought to improve the stock of farmers of small means in Wales was shown to advantage at the Abergele show in 1936 where, in the Hereford cattle section, the champion beast was none other than the Breconshire premium bull Leen Generosity, in the care of T. E. Gwillim of Ffostill, Talgarth. At the same time, we have noted that smaller farmers were more likely to exhibit in the local or district classes.

22. Leen Generosity, champion Hereford at the 1936 show.

   Doubtless encouraged by David Davies, from 1922 onwards the Society sought to foster improvements in the breed of Welsh foxhounds. Davies's commitment was to be seen in his inaugurating the National Kennel Stud Book at the 1922 Wrexham show, a much-needed venture in view of the deterioration of the old breed during the war years when it had been difficult to feed the packs because of food restrictions and when the huntsmen were away. The Wrexham show's last day, 28 July 1922, also witnessed a new competition for foxhounds, Welsh, cross-breds and English, with a hundred couples being colourfully paraded by huntsmen watched by a huge crowd. The start of an important feature of the show during the inter-war years, it proved to be a popular crowd-puller, not least among the ladies of fashion of well-known county families, as well as demonstrating the remarkable improvement that quickly occurred in Welsh and cross-bred hounds under the stimulus of the Welsh Foxhound Stud Book. However, perhaps because of financial exigencies, only three hound shows were held between 1929 and 1936, one at the 1931 show. On this last occasion, David Davies, disappointed at the calling off of his pet inter-county Welsh native breed competitions at the Llanelli show because of want of sufficient entries, was consoled by the finest turnout of hounds ever staged at a Royal Welsh Show and by Verity securing

23. Welsh hounds in the parade at the Swansea show, 1927.

the coveted place of champion Welsh bitch for his pack. From 1936, less cordial relations prevailed between the Royal Welsh Agricultural Society and the Welsh Hound Association, formed in 1922, and thereafter – unlike the earlier cooperation – any hound show put on at the Royal Welsh showground was to be organized and managed solely by the Welsh Hound Association.

One of the side shows at the Royal Welsh stretching from pre-war days was the dog section, which, though confined to a single day, was to become an increasingly popular feature of the event, especially when the show was held in industrialized urban areas. Thus the dog section at the Welsh National Agricultural Society's show at Newport in 1914 attracted no fewer than 1,200 entries. Bridgend's dog classes in 1924 were the strongest yet, reflecting the growth in popularity of dogs in south-east Wales. What was described as 'the cult of the Alsatian', which had been growing in the south-east for some years but which received a big impetus by the recent visit of the famous French and Belgian police dogs to Cardiff, was mirrored in the keen interest the dogs excited among the Bridgend crowd. The open dog show at Wrexham's Royal Welsh in 1928 again proved to be very successful by virtue of its attracting huge crowds from the industrial areas, who came to watch some of the leading dogs from across the British Isles compete in their different classes. Ironically, however, when the dog show of 1928 made a loss, the Society, faced with the need to economize, eliminated the dog section from its 1929 schedule. In response to the Haverfordwest local committee's pressing for a dog section to be reinstated at the forthcoming show, a seemingly successful dog show was staged there in 1935. It was, however, discontinued for the rest of this period.

Other side shows were those of forestry, poultry and horticulture, and, like the dog section, these, in the face of the need to reduce show expenditure, came under threat after the 1920s. Both forestry and poultry sections incurred losses from 1928 onwards. Crisis point was reached by 1931 when Council was forced to decide whether the forestry or poultry sections should be omitted from the 1932 schedule. Forestry, we saw earlier, was the casualty and was not restored until 1936. That should in no way suggest, however, that this section had been sub-standard. Quite the contrary, for, after beginning in a comparatively small way in 1923, great strides were made in its quality in the mid-1920s when Osmond Smith, the Glamorgan county land agent, had taken charge; his influence effected a marked improvement at the Bridgend show in 1924 and an even better display at Carmarthen the following year when films on forestry were shown in the cinema erected in the forestry enclosure. Importantly, here we see the Royal Welsh Agricultural Society reacting to recent developments in forestry in Wales – just as it was to do in relation to those in dairying – by giving it adequate attention in the arrangements and programme of its show. Allegedly the somewhat indifferent reinstated forestry exhibition at the Abergele show in 1936 suffered from the want of advice and help from the late Osmond Smith. Nevertheless, forestry at the Caernarfon show in 1939 was said to have made a strong appeal: one of its exhibitors was 76-year-old James Walters of Llwynfedwyn, Farmers, near Lampeter, who was awarded two silver championship medals for craftwork and basket-making. A lifelong farmer, he had been making these things as a hobby since he was fourteen.

Recognized by a press commentator in 1926 as playing a part in the improvement of poultry farming – as well as pigeon breeding – in Wales, the poultry show was to survive intact from its pre-war days until 1939. Just as the produce department of the show was to make great headway after E. P. Norton took over its management from 1930, the poultry section was to benefit from the devotion of H. H. Perry of Llandogo, near Chepstow, who, likewise from 1930, became honorary poultry show manager. Typical of so many stalwarts of his day, he gave unselfishly of his energies to the point that, when he was far from well, he insisted on carrying out his duties at Llandudno in 1934, only to die within a fortnight.

The horticultural section, too, made a loss at the 1928 show, despite its popularity on that occasion as at earlier shows. This meant that a separate charge was made for admission to the tent the following year and its very survival as a show section hung in the balance. The Caernarfonshire county meeting for 1933 thus viewed with dismay the possibility of the horticultural section being omitted from the 1934 show schedule. At the same time, criticisms of the standard of the horticulture section were aired at the 1933 county meetings of both Cardiganshire and Breconshire, the former stating bluntly that the competitive classes were unworthy of an exhibition of national standing and urging the Council to set up a permanent horticulture sub-committee as a means of improving the classes. It is not surprising, therefore, that a revision of the horticultural section occurred in the 1937 show and in many respects it was overdue. For, by virtue of the fact that payment of a small lump sum had hitherto entitled a prospective exhibitor to enter each of the multitude of classes throughout the entire schedule, last-minute disorgan-ization of the layout of a large part of the horticultural tent ensued through visitors staging exhibits only within a small number of classes but having neglected to inform the organizers of this. The schedule for the 1937 show was entirely revised, so far as the open section was concerned. While the special class for trade displays was retained, the classes for small exhibits were omitted in favour of substantial exhibits composed of groups of miscellaneous plants and collections of hardy perennial plants and cut blooms and also of sweet peas, cut roses and vegetables.

The timbering and horticultural sections were only part of a much wider programme mounted at the annual show with the aim of educating Welsh agriculturalists and rural dwellers in the most up-to-date farming techniques and appliances. In placing high priority upon this educational aspect to the show the Society was, of course, carrying on from the lead given in this direction at Aberystwyth. Nevertheless, the migratory shows between 1910 and 1939 witnessed an increase in the scale and tempo of educational exhibits relating to progressive farming methods which was in itself a reflection of the quickening pace of the application of science – mechanical and chemical – at this time to the problems of the farm and rural life. For the *Western Mail* reporter on the 1927 show the exhibits by the trade stands and educational bodies at successive Royal Welsh Shows could match those of any show of comparable size in the United Kingdom. Indeed, Captain Howson emphasized in the early 1930s that the Society considered that the educational aspect of the annual show was of first importance, and that the Royal Welsh Show had always been an educational one. Among the various public bodies and institutions – over and beyond the private firms – which submitted exhibits of one kind or another at the various Royal Welsh Shows were the English and Welsh departments of the Ministry of Agriculture and Fisheries, the Agricultural Committees and Education Authorities of the Welsh county councils, the various county federations of the Women's Institutes, His Majesty's Forestry Commission, the Welsh Agricultural Organisation Society, the Rural Industries Bureau, and, importantly, the various departments of the University Colleges at Bangor and Aberystwyth.

Mention needs to be made of some of the more important exhibits during this phase of the show's existence. At the shows of the late 1920s and 1930s as, indeed, at all national and county shows within the United Kingdom, visitors had an opportunity to see the marketing demonstrations of the Ministry of Agriculture which sought to bring improved methods of marketing to the notice of producers and distributors of home produce; throughout, emphasis was placed on the importance of standardization and on the value of the national mark which was being applied to various commodities in turn. Thus the Aberystwyth show in 1933 saw the Ministry of Agriculture stage egg-grading demonstrations in cooperation with the Clunderwen egg-packing station with the aim of arousing still further interest among farmers in the sale of eggs under the national mark grades. By January 1939 Council felt that these marketing demonstrations had served their purpose, and so they were replaced at the Caernarfon show in the following summer by a general agricultural education and marketing exhibition, organized by the Ministry of Agriculture with the assistance of the School of Agriculture at the University College of North Wales, Bangor, and other interested public bodies. Likewise, the Ministry of Agriculture mounted annual exhibits from the 1923 show onwards, demonstrating the means and appliances for the production of clean milk, significantly a subject that was receiving special attention in Wales at this time and which mirrored the trend towards the production of Welsh breeds of cattle for dairy purposes. Liquid milk production, unlike so many other food products marketed by British farmers in these inter-war years, was not hit by fierce overseas competition. As part of the 'Drink More Milk Campaign' promoted in the 1930s by the newly founded Milk Marketing Board (1933) and the National Milk Publicity Council, in 1935 the Society introduced a milk bar at its showground at Haverfordwest. Toasts at the show's luncheon that year were drunk in milk which, so a *Western Mail* editorial claimed, meant there was a noticeable absence of that 'wild and woolly rhetoric' which was sometimes heard at agricultural and other dinners when the toasts were drunk in stronger stuff. By January 1937 the Society's officials believed that, in view of the general improvement in the conditions under which milk was produced, it was no longer desirable to continue demonstrations in the production of clean milk at its shows. Instead, competitions for milk produced locally were held at the Monmouth show later that year.

Apart from these Ministry of Agriculture exhibits, there were a host of others such as crop-drying demonstrations at Carmarthen in 1925 during which harvested wet crops were dried out with forced air; sheep-dipping demonstrations by Messrs Cooper, McDougall and Robertson Ltd, at the Llandudno show in 1934; regular demonstrations in butter-making; exhibits, from the 1929 show onwards, of the rural industries of Wales, including in that year a spinning demonstration in angora wool, the first time spinning had been demonstrated at the show; regular exhibitions by the National Federation of Women's Institutes – the latter playing an increasingly important and valuable role in the shows of the inter-war period – which demonstrated a variety of farm, kitchen and garden produce and examples of needlework, knitting, furcraft, rugs, quilting and smocking; and, as a final example, many exhibits over the years by the various departments of the University Colleges of Wales, such as that put on at Abergele in 1936 by the School of Agriculture at the University College at Bangor which embraced animal husbandry, animal diseases, dairy bacteriology, plant pathology, agricultural zoology, the North Wales Seed Potato Scheme, a soil survey of Wales, and other subjects.

An increasingly striking and extensive feature of the shows between 1910 and 1939 was the trade section, whose stalls and stands not only provided the opportunity for much business but also served as a means of education and instruction. In their yearly pilgrimage to the show, Welsh rural dwellers were afforded the opportunity, which otherwise might never have come their way, of keeping in close touch with those advances which were continually being made in machinery and implements, manures, seeds and other articles of use in their daily lives. Above all, these were the years when the motor engine came into its own in rural Britain, in the process sadly spelling the eventual demise of the horse. The shows during those years saw displays by various firms, both Welsh and English, of fertilizers, animal and poultry feed, hay mowers, horse rakes, fertilizer distributors, Austin and Fordson tractors, livestock trucks, like the Chevrolet vehicles displayed at Cardiff in 1929, lorries, and vans such as the Austin delivery vans and Commer commercial vehicles on show at Haverfordwest in 1935. All were symbolic of the evolutionary process – rapidly forging ahead in the late 1920s – of the mechanization of the Welsh farmstead. Albeit, such was the horse's lasting sway on Welsh holdings that as late as 1939 there were only 1,932 tractors in Wales.

From its Aberystwyth days the show had staged events like harness and jumping competitions and farmers' and tradesmen's turnouts to pull in the crowds. During the period between 1910 and 1939 these entertainments in the form of ring competitions and displays were to continue in ever more varied and lavish fashion. Shows in these years came to include attractions like horse leaping and equitation classes, driving, trotting handicaps, and, ever popular with showgoers, children's riding pony classes. Surely unique in these latter classes at any Royal Welsh Show ever held were the five sisters aged between seven and fourteen, daughters of the noted Montgomeryshire riding enthusiast H. Meyrick Jones, who competed against one another at the 1937 Monmouth show! Skilled horsemanship was also displayed by the military, as at the Newport show in 1914 when a very attractive item was the musical ride and charge given by His Majesty's Dragoon Guards. Following the exemption of agricultural shows employing the services of musicians from the entertainment tax – a concession which was itself largely the response to efforts made by the Royal Welsh Agricultural Society – the Royal Welsh was enlivened by the presence of a military band, the first at Welshpool in 1923 of the 2nd Battalion of the Welch Regiment. Colour was added and visitors attracted, too, by the masters of local packs parading their foxhounds in the ring. This occurred independently of the Welsh Hound Association show, which took place, we have seen, at certain of the Royal Welsh Shows. One important new attraction on the showground from 1925 onwards were the sheep dog trials. Their introduction immediately boosted the attendance during the last day at both Carmarthen (1925) and Bangor (1926). Competing for the 50-guinea silver challenge cup presented in the Royal Welsh Stakes by the generous patron and chairman, David Davies, these trials were keenly contested. Rapt crowds were thrilled and sometimes spellbound by the almost uncanny feats of skill of some of the finest sheepdogs in the kingdom. Engendering rivalry between northern and southern parts of Wales, at Haverfordwest in 1935 the cup was won for the third time by John Jones of Trawsfynydd and so became his property. The inclusion of these trials was a

24. John Jones with his sheepdog Blackie and the Challenge Cup, presented by Lord Davies, which he won outright in 'the Royal Welsh Stakes' at Haverfordwest, 1935.

valued and distinctive feature, and Welsh visitors to the Royal Show held at Cardiff in 1938 lamented the fact that no opportunity was afforded the Welsh shepherd to display his skills with his faithful sheepdog.

In the early 1930s various county meetings called for the introduction of more spectacular events into the show programme in order to attract visitors and, in particular, that the afternoon and evening ring events of the first two days and those of the morning of the third day be made more interesting and attractive to the ordinary visitor. Bowing to this pressure, Council dispensed with the services of a band at the Haverfordwest show in 1935, substituting ring events of a spectacular nature which took the form of daily displays of horsemanship by non-commissioned officers from the School of Equitation at Weedon. Approving this, the county meetings of Glamorgan, Denbighshire and Flintshire recommended the inclusion of similar ring events at Abergele in the following summer and at all future shows. Visitors to Abergele were accordingly entertained by the combined horse and motorcycle displays given by the non-commissioned officers and men of the Royal Corps of Signals. While recognizing in 1935 that spectacular features entailed heavy expenditure, David Davies acknowledged that they appeared at the time to form a regular item in the programme of all the major agricultural shows, and that most show officials viewed them as a necessary means of attracting the public. A warning note was nevertheless sounded by the Pembrokeshire county meeting of 1936 in its request that the Council consider whether this tendency towards increasing extraneous attractions, leading as it did many visitors to spend a lot of their time in the vicinity of the main ring, might not ultimately result in a reduction of the support accorded to agricultural shows by trade firms. The following year, the kind of sentiment expressed in Montgomeryshire's call to make the afternoon programme in the main ring more of an entertainment than an exhibition was to concern C. Bryner Jones, who was firmly wedded to the ideal of the show providing demonstrations.

Certainly many of the county families, besides valuing the show as an agricultural shop window, looked upon it as a social occasion where they could be noticed and mix with the best-known gentry and aristocratic families from neighbouring shires and further afield. Prominent local families entertained house parties at their mansions during show week and gave small luncheon parties on the showground. Not least in these polite circles were the masters of foxhounds and their wives. Genteel ladies dressed in the latest fashions, their toilettes carefully described in the columns of the *Western Mail* and no doubt in local newspapers. As well as being drawn towards watching the hound show, certain of these ladies rode in the hunter class and especially the lightweight classes. At the Carmarthen show of 1925, for instance, Lady Kylsant, wife of the president of the Society, had three of her grown-up daughters in the riding ring, while at a wet Swansea show two years later a prominent lady rider was the Honourable Anne Lewis of Hean Castle, Pembrokeshire, daughter of Lord Merthyr who was winner at that same show of the class of Shorthorn bulls calved before 1924 with Hean Arthur, recently placed second at the Royal.

It will be apparent that the show in these years was a developing institution, with many innovations giving it an even more comprehensive, valuable and interesting air. New features like the ambulance competitions (1912), hounds section (1922), clean milk demonstration and forestry classes (both in 1923), Inter-County Group and Breed Competitions (1924), the cinema and sheep-dog trials (both in 1925), the Welsh Rural Industries and Crafts exhibit (1929), the Ministry of Agriculture's exhibition of grading and marketing (from the close of the 1920s), the wool class (1932), and the agricultural education and marketing exhibition (1939) all bore testimony to the Society's concern to bring scientific, technical and commercial knowledge to the Welsh country-side. Of course, this was the educational ideal and just how many farmers modernized as a result of their visits is impossible to measure. (Perhaps not too many of them if we accept Moore-Colyer's claim that Welsh farmers in the inter-war years 'appear to have been reluctant to invest in innovative measures'.) Importantly, the Society began targeting young farmers in order to achieve progress. In 1926 at Bangor came yet one more new event with the introduction of the amateur judging competition open to farmers' sons under twenty-five years of age, who were put to judge livestock. By the 1937 show, entries to the competition were the largest on record. Both the Merioneth and Breconshire county meetings for 1931 were alive to the importance of young farmers being drawn into the Society's membership, which led to the new ruling in late 1932 that members of recognized Young Farmers' Clubs and students of farm institutes and colleges, not engaged in farming on their own account, be accepted as members of the Society on pay-ment of a reduced annual subscription of 10s. 6d. This was, however, simply the prelude to their later high profile at the Society's shows in the years following the Second World War.

Apart from introducing new features into its shows, the Royal Welsh Agricultural Society also sought to improve facilities on the showground. There can be little doubt that the Council endeavoured to heed the calls made by county meetings in 1931 and 1935 for improved standards of stewarding. At the Haverfordwest show of 1935, cheaper food was provided for the general public, who might not have been able to afford the prices charged in the first- and second-class refreshment marquees. Also urgent was the need to improve the conditions of those described by E. Verley Merchant in 1936 as the 'unseen army' of herdsmen. Unsatisfactory provision for their comfort and enter-tainment was a talking point until the late 1930s. Some recognition of the need to treat them well was made when, following the practice of a number of other leading societies, the Council decided to entertain stockmen and others employed at the 1939 Caernarfon show to a supper and smoking concert on the second evening. Attended by 280 persons it was a great success and doubtless helped engender the all-important good feeling between officials and stockmen.

A more extensive publicizing of the show came about over these years, too, through a mixture of traditional and new means of communication. Particularly strong in its support of the Royal Welsh Show was the *Daily Mail*, which, for instance, appointed its special correspondent I. D. W. Izzard to cover the Cardiff show in 1929 and also sent its loudspeaker van to shows in the 1930s free of charge. Faithful in its coverage, too, from the beginning was the *Western Mail*, with valuable informed show comment being supplied by correspondents like Walford Lloyd and Charles E. Lloyd in the 1920s, followed by W. H. Woodcock in the 1930s. Such support led the paper to request, year after year, stand space at the show free of charge but to no avail, the Society arguing it would be unfair on others to grant this favour unilaterally. The *Liverpool Post* also carried full reports. Moreover, the shows were covered by agricultural journals, like the *Livestock Journal* and *Farmer and Stockbreeder*. The first film of a Royal Welsh Show was taken at Llanelli in 1931 by Universal Pictures Ltd, which the Society subsequently purchased for £12 10s. For its part, Movietone News was granted the sole right to take a sound film of the Llandrindod Wells show the following year; taken free of charge, it was distributed to 75 per cent of the country's cinemas. That such films were used for publicity purposes can be seen by the acceptance of an offer in 1933 from Messrs Cooper, McDougall and Robertson Ltd to exhibit the Society's show film, in conjunction with one of their own, during tours of Breconshire and the Llanrwst district of Denbighshire, arranged by them in conjunction with the National Farmers' Union. A new feature, too, were the wireless broadcasts from 1929 onwards (if not before), although as yet BBC coverage seems to have been largely confined to half-an-hour's 'talk' on one of the evenings before or during the show.

This period also witnessed the beginning of visits to the show of people from overseas, which were to become an important feature over the years. The Council was thus informed by the secretary at the 1927 show that steps had been taken to ensure the attendance of foreign visitors at future shows by securing Wales's inclusion in the future itineraries of such visitors. Wrexham's show the following year accordingly received a visit by the Empire farmers who were touring the British Isles. Not only was the cream of Welsh livestock thereby on show to overseas farmers, but a coloured film of the Caernarfon show of 1939 was taken for the Canadian Pacific Railway Company at the request of the Montreal office which, Verley Merchant surmised, might stimulate the demand for home-bred cattle.

CHAPTER SIX

Tubercle-free Herds and Wild Rabbits

In addition to staging its annual show, the Society in these years continued to promote activities and campaigns which it felt served the interests of Welsh farmers. At the beginning of 1933 a joint-committee – composed of representatives of the King Edward VII Welsh National Memorial Association and the Royal Welsh Agricultural Society – was set up for the purpose of encouraging the establishment of tubercle-free herds of cattle. Representing the Society were Lieutenant Colonel G. E. FitzHugh of Plas Power, Wrexham, a Council member, Moses Griffith of Devil's Bridge, Cardiganshire, then lands' director under the Cahn Hill Improvement Scheme and a member of the Society's Finance and Executive Committee, and T. A. Howson, the secretary. They were to take part in the first joint-committee meeting held at Shrewsbury on 2 February 1933. Additional members later nominated by the Society to the joint-committee were three academics from the University Colleges at Cardiff and Bangor and G. Herbert Llewellin, of the renowned milk-churn manufacturing family of Haverfordwest and a member of the Council. In due course, the joint-committee presented to the Councils of the Association and the Society respectively a very wide-ranging report, upon receipt of which the Council of the Royal Welsh Agricultural Society responded by appointing a sub-committee to consider it and to submit recommendations regarding what action the Society might usefully take. Sitting on the sub-committee were the Society's representatives on the joint-committee and, in addition, Thomas Evans, a trustee and Council member, and Professor R. G. White of the University College of North Wales, also a Council member. It met at Shrewsbury on 21 November 1934 and made recommendations. Theirs was a tricky balance to get right; on the one hand they endeavoured to support any recommendations of the joint-committee which might encourage the provision of an adequate supply of pure milk and boost public confidence in it while, on the other, they were careful to avoid advocating support of those recommendations of a highly controversial nature or any which, if adopted, might defeat the primary function of the Society which was the advancement of the interests of the agricultural community. Accordingly, they decided that they could not advise the Council to support certain of the recommendations advanced by the joint-committee.

The sub-committee adopted the joint-committee's recommendation to undertake a propaganda campaign to educate the general public in the value of liquid milk as a food and to stimulate the demand for it by lectures to women's institutes and schools and joint exhibitions of the Welsh National Memorial Association and the Royal Welsh Agricultural Society at the annual Royal Welsh Show and at other agricultural shows. Anxious to protect farmers' profits, however, the sub-committee made a proviso that in the event of efforts made to stage an exhibit at future Royal Welsh Shows, care should be taken to avoid staging anything which might tend to create a 'milk scare'. It went on to adopt some, though not all, of the joint-committee's recommendations towards ensuring that the milk supplied should be of good hygienic quality and conform to a definite standard as to its freedom from tubercle bacilli. These included, in the first place, that local authorities should be obliged to appoint full-time veterinary officers to carry out inspections and give free advice to farmers in the method of prevention and eradication of tuberculosis, and that milk from all herds kept for supplying the public should be biologically examined for tubercle bacilli at least twice a year, and, where the results were unsatisfactory, effective steps be taken to trace and eliminate the source of infection. Also adopted by the sub-committee were the recommendations that measures should be taken by the Royal Welsh Agricultural Society and the agricultural education committees of the county councils of Wales and Monmouthshire to bring to the notice of Welsh farmers the possibility of utilizing the 'Bang' method of eradicating tuberculosis from herds (that is, by separating tuberculin-negative and tuberculin-positive stock) and that county councils should explore the practicability of assisting farmers in the segregation of reacting animals. A final recommendation adopted was that a scheme should be initiated to create a tubercle-free dairy cattle area in Wales, perhaps in Anglesey, in order to prove its feasibility and the desirability of its extension, the sub-committee suggesting that the Ministries of Health and Agriculture be urged to explore the possibility of starting such a scheme in Wales. Members of the sub-committee recognized that the adoption of many of these recommendations would impose a very heavy burden on those counties in which milk was mainly produced for consumption elsewhere, and they thus urged that the government provide grants to encourage such counties to adopt and implement the recommendations.

After long deliberation a special meeting of the Council held on 18 March 1935 adopted the sub-committee's recommendations. An enthusiastic Lord Davies argued that the rigid application of these measures 'would mark a very decided first step towards the, for all practical purposes, total abolition of a terrible scourge from our cattle and thereby reduce, very materially, the general incidence of tuberculosis in human beings'. In so far as the recommendation for the creation of tubercle-free breeding areas in Wales was concerned, a prompt response came in a survey made by Dr R. F. Montgomerie, veterinary adviser to the School of Agriculture at University College of North Wales, Bangor, and his colleague W. T. Rowlands, of the cattle in certain upland rearing districts in north Wales. Their conclusion that the establishment of extensive disease-free areas was a feasible proposition was extremely encouraging.

The Society went on to take a number of initiatives in the attempt to establish tubercle-free herds. At the special meeting of 18 March the secretary was instructed to convey to the Ministry of Agriculture and Fisheries that Council wished to associate itself with the communication addressed to the minister on 6 February by the Council of the Royal Agricultural Society of England, drawing his attention to the shortcomings of the Tuberculosis (Attested Herds) Scheme. Moreover, the meeting invited G. H. Llewellin to prepare a paper suggesting amendments to the Ministry's Attested Herds Scheme which might, if adopted, render it a practicable and useful measure. The meeting also requested the chairman (Lord Davies), Baron de Rutzen of Slebech Park, Pembrokeshire, G. H. Llewellin and the secretary to prepare and forward to the minister of agriculture and fisheries a resolution directing his attention to the increasing efforts being made throughout the United Kingdom to free herds from tuberculosis and to provide a safe milk supply, and urging him to take action to ensure that, in future, all cattle imported from Ireland or other overseas countries 'for breeding or dairying purposes' should be certified as having passed the tuberculin test prior to shipment. Such a resolution was sent to the minister in due course.

As hinted, the Tuberculosis (Attested Herds) Scheme which became operative in February 1935 had met with some objection. In response to the news that the Ministry was contemplating making amendments to its scheme, G. H. Llewellin requested the Society's Council meeting on 24 July 1935 to defer consideration of his memorandum on the scheme (copies of which had been circulated) pending the announcement of the amendments which the Ministry had in mind. His wish granted, a sub-committee met in early September 1935 to consider the revised Tuberculosis (Attested Herds) Scheme, which had been submitted in August to the various agricultural bodies of the kingdom. Their deliberations led to recommendations being sent back for the Ministry's consideration. Surrounded by such controversy and criticism, it is hardly surprising that the scheme had not met with any marked enthusiasm in England by the close of 1936 and that it was practically ignored in north Wales. On the other hand, it was gratifying that no fewer than 42 of the 117 English and Welsh herds which had been attested up to 31 July 1936 were to be found in Carmarthenshire and Pembrokeshire. Noting that only three of the attested herds were from north Wales, Lord Davies urged farmers there to get their herds attested the following year.

Another problem afflicting the rural community, which the Royal Welsh Agricultural Society addressed in the late 1930s, was the wild rabbit menace. So widespread was the nuisance that a Select Committee of the House of Lords on Agriculture (Damage by Rabbits) in 1936–7 heard evidence, including the extent to which Skokholm Island off the coast of Pembrokeshire was affected. That the Society would play a major part in attempting to counter the problem was made clear in its response to a letter received on 30 June 1938 from R. Williams-Ellis, Glasfryn, Chwilog, Caernarfonshire, chairman of the Council of the Welsh Agricultural Organisation Society, in which he expressed the opinion that the Royal Welsh Agricultural Society would be the best and most effective body to initiate action in Wales, since it was more fully representative of the landowner and tenant interests in this particular matter than any other body. Accordingly, the Council meeting of 6 July 1938, recognizing the enormous amount of damage caused annually by wild rabbits and the desirability of effecting an immediate and drastic reduction in their numbers, appointed a sub-committee – which included C. Bryner Jones and D. D. Williams – to consider the steps which the Society might take towards assisting in the mitigation or abolition of the wild rabbit menace. It was further resolved that, prior to the meeting of the sub-committee, the secretary should communicate with the National Farmers' Union, the Country Landowners' Association and like bodies to ascertain what, if any, action they had taken in relation to the problem.

A meeting of the sub-committee on 7 October 1938 heard evidence of the seriousness of the problem, particularly in Pembrokeshire, Cardiganshire and Caernarfonshire. According to Captain W. H. Buckley of Castell Gorfod, St Clears, the problem in south-west Wales was partly caused by the 'rabbit industry' operated by trappers and dealers, which had grown to such proportions that 'now light lorries laden with rabbits destined for the large towns roared through the countryside at night'. The price of rabbits on the farm was about 6d. each, of which 3d. went to the farmer and 3d. to the trapper. In response to recommendations made by the sub-committee, the Finance and Executive Committee later that October recommended that Council adopt the following resolutions:

i) that the Minister of Agriculture and Fisheries take immediate steps to promote legislation with the object of (a) placing upon the landowner and occupier of land the onus of keeping wild rabbits within such limits as to numbers or under such control, as will render them innocuous to neighbours and (b) enabling a person whose property is injured by wild rabbits to claim compensation from any adjoining landowner and occupier who fail to comply with any order which may have been properly made for the reduction or control of the rabbits on the land in their ownership and occupation.

ii) that, in the opinion of the Sub-Committee, the Minister should be authorised to delegate to Local Authorities the execution, through their Agricultural Committees, of any powers for combating the wild rabbit menace which may be conferred upon him by legislation.

Council adopted these recommendations at its meeting on 1 December 1938.

In the late 1930s, pressure was also exerted on the minister of agriculture in order to achieve other goals. One such approach came with the Society's decision at the close of 1938 to support the representations made by the National Farmers' Union about the current disastrous position of the sheep industry in the country and to urge the government to take immediate steps to alleviate the problem. A later approach was made towards the close of 1939, when the Society protested over the 'short-sighted and retrograde' policy of withdrawing the grants which had hitherto been made under the Livestock Improvement Scheme for the provision of premiums for approved stallions, bulls, rams, and boars and the grants in aid of milk recording. Accordingly, the minister was urged to take steps at the earliest opportunity to revive these grants, which were perceived to have been unwisely withdrawn under the exigencies of war. Prompted by Moses Griffith, a meeting of the Finance and Executive Committee of early October 1939 decided to recommend Council to submit yet another representation to the minister of agriculture calling for the lifting of the recent suspension of the Licensing of Bulls Act, 1931. The continuation of such a suspension, it was felt, would seriously harm the cattle-breeding industry of Wales, for, as a store-raising country, the licensing of bulls was vital if commercial store cattle bred in Wales were to maintain their recent improvement and not deteriorate – to the advantage of Irish stores, long protected by licensing of bulls – by once again giving a free rein to the old 'scrub' bull. Fortunately, the Ministry's decision to withdraw the suspension of the 1931 Act rendered the Council's representation to the minister unnecessary.

From the close of the 1920s the Society extended its responsibilities by shouldering the secretarial work of some of the Welsh breed societies, with which close relations had evolved from early days. In late November 1927 Colonel David Davies maintained that it would be to the mutual advantage of the Royal Welsh Agricultural Society and of all the Welsh breed societies if the Royal Welsh staff carried out the secretarial and administrative work of the breed societies. Thereby the Stud, Herd or Flock Books issued by the national breed societies would in due course be compiled by an expert staff, and, furthermore, properly organized schemes of advertising and propaganda could be formulated and put into operation with the minimum of expense to these societies by virtue of their cooperating for this purpose and utilizing the Royal Welsh Agricultural Society as their agent. For its part, the Royal Welsh Agricultural Society would gain both directly and indirectly inasmuch as its continued success depended in no small measure on the advancement of Wales's national breeds. The outcome of this offer, understandably facilitated by its honorary secretary, Captain Howson, who was closely involved in the affairs of the Welsh breed societies, was that from 1928 or 1929 onwards the secretarial work of the Welsh Pony and Cob Society, the Welsh Pig Society and the Welsh Hound Association (formed in 1922) was carried out by the staff of the Royal Welsh Agricultural Society for an agreed payment, which in the case of the first two societies mentioned was reduced to £30 a year from 1932 onwards to take account of the difficult times they were facing. A disagreement over fees paid to the Royal Welsh Agricultural Society led to the termination of the arrangement with the Welsh Hound Association in 1937. Of course, these additional secretarial duties placed extra pressure on the office staff; such was the increased burden shouldered by Captain Howson, that it led to some friction between him and the show's honorary director, Reuben Haigh, who complained in late 1929 that attending meetings of the breed societies in show week precluded the secretary from giving sufficient time to the management of the show itself.

We saw in Chapter Three that production of the Society's *Journal* lapsed from 1909 until 1923. Upon its relaunch a one-year publication was issued, styled the *Annual Report*, but from 1928 this was renamed *The Journal of the Royal Welsh Agricultural Society*. It was to continue uninterrupted until 1939, although Isaac Jones, of Llysfasi Farm Institute, Ruthin, was to forward a motion at the close of 1932 that the *Journal* should be discontinued temporarily rather than sections of the show be abolished because of the difficult times. Fortunately, nothing came of this proposal. One significant development regarding its circulation came in late 1930 when it was decided that, in accordance with the practice of other similar organizations, the *Journal* should henceforth be sold to the general public for 5s. a copy.

As well as carrying a lengthy review of the Society's show and providing particulars of its financial standing and the names of its officers and members, successive issues contained articles on developments in Welsh agriculture, thus informing the rural community of scientific and marketing advances in the industry. In keeping with their role as researchers within the field of Welsh agricultural science and economics, members of staff at the University Colleges at Aberystwyth and Bangor were prominent contributors of notes or articles to the *Journal*: Professor R. G. Stapledon, as its director, and, in 1937 and 1938, Dr T. J. Jenkin updated farmers on the work of the Welsh Plant Breeding Station at Aberystwyth; Professor R. G. White of Bangor wrote (in 1929) on the 'Control of the Warble Fly' and (in 1931) on 'The Importance of Hardiness in Welsh Mountain Sheep'; various articles were written on the three marketing schemes – milk, pigs and potatoes – introduced by the government in 1933, including one (in 1933) entitled 'The Birth of a New Agriculture' by the aforementioned Isaac Jones of Llysfasi Farm Institute, and another the following year by A. W. Ashby, W. H. Jones and J. R. E. Phillips, of the Department of Agricultural Economics at Aberystwyth, in which they argued that the Milk Marketing Scheme regulating the marketing of milk was the one most likely to make a difference to Welsh farming, indeed 'the only one that at the moment appears likely to exercise any important influences in the development of new agriculture'. Other articles kept members and the public informed of the incidence of tuberculosis in cattle-breeding districts and of the efforts being made in the 1930s to encourage the establishment of tubercle-free herds of cattle.

Lord Davies was to explain in his *Report of the Show and the General Working of the Society* for the year 1939 how, with the onset of war, the decision had been made to bring out a *Journal* for that year only after long discussion by the Council. What swayed the balance in favour of publication was the consideration of the importance of preserving continuity of essential records in a form which had now become familiar to members. Even so, to effect economies it was decided to dispense with the usual full review of the summer show and to omit the customary articles. Some 1,500 copies of the *Journal* were produced in 1939.

Improvements effected in the running of the Society such as the establishment of county committees, the growth of the show and the Society's increasing involvement in the wider agricultural domain were to allow Lord Davies to reflect at the close of 1939:

We can, I think, look back with justifiable pride upon the steady progress made by the Society since its resuscitation in 1922 . . . Slowly but surely we have extended our prestige and sphere of influence until the Society has assumed a position of real importance in the world of agriculture, while the well-known names which figured in the catalogue of last summer's meeting at Caernarvon are a sufficient indication of the esteem in which our shows are held.

After the lapse in its activities for the duration of the Second World War, new challenges would be presented by the changes wrought in the farming industry and by the mounting costs of staging an annual show.

PART THREE

Deficits and Mud

THE SECOND MIGRATORY PHASE, 1947–1962

CHAPTER SEVEN

A New Headache

MOUNTING COSTS ALL ROUND

Soon after the outbreak of war in 1939, the Council on 20 November *The war years* decided that the Society's activities and expenditure should be reduced to the minimum compatible with keeping the organization in existence, conducting routine business which might arise from time to time and avoiding as far as possible the infliction of undue hardship upon members of staff. An Emergency Committee, composed of the chairman and six others, was appointed and empowered to conduct affairs until the termination of hostilities, or during any other period determined by the Council. It was also decided that the annual show should be discontinued until the Council, on the recommendation of the Emergency Committee, directed and that no further meetings of the Council or standing committees would be held until the Emergency Committee should so decide. Given that these recommendations would absorb more than the current income from invested funds, members were urged to continue to pay their subscriptions so that the Society would be in as favourable a position as possible when the time came to resume its activities. On restarting, the present members of staff were to be reinstated as far as possible. Finally, the Emergency Committee was to review the Society's position within twelve months. Tenancy of the offices in Queen Street, Wrexham, was terminated and the Society moved to rent-free office and storage accommodation in Ruabon, thanks to the generosity of Reuben Haigh who placed his business premises there at the Society's disposal for the duration of the war.

That thought was being given to the Society's post-war future was plainly in evidence at a meeting of the Emergency Committee in August 1944. Speaking from the chair, now that Lord Davies had passed away earlier in the year, C. Bryner Jones suggested that future shows ought to be developed more on the lines of demonstrations than sporting events. Moses Griffith urged that the sooner Council obtained the views of the various interests concerned and drew up programmes for its future activities the better the position of the Society would be upon its recommencement after the war. As far as shows were concerned, there was no doubt, he averred, that new ideas would have to be adopted. In his view there was a wide field for innovations in the organizing of shows, especially in Wales. For instance, consideration would need to be given to the desirability or otherwise of continuing to schedule classes for young cattle, to the question of making provision for progeny tests, and to the desirability of catering for new movements by Young Farmers' Clubs. He believed that the appointment by the Emergency Committee of a small sub-committee would help to thrash matters out and formulate a concrete proposal for consideration by Council. Reflecting a certain amount of feeling among the members, the honorary director, Reuben Haigh, suggested that, were a sub-committee to be formed, it might be advisable to introduce 'a little new and young blood into its composition'.

Much thought was obviously being given to the future both by Council members and by those present at county meetings. At a Council meeting held in Morris's Café, Shrewsbury, on 20 November 1945 a sub-committee was finally appointed to consider the future organization and management of the annual show and the secretary agreed to attend a meeting of the Provincial Publicity Committee at Aberystwyth on 25 April 1946 for the purpose of discussing methods of improving future shows. County meetings held in 1946 were restless, and motions were forwarded to the Council on a variety of issues. Caernarfonshire called for records of attendance at Council meetings to be kept in future and that those who failed to attend a reasonable number during the year be removed from office; Denbighshire and Flintshire, echoing Reuben Haigh, requested that efforts be made to infuse more 'young blood' into the Council; another communication from Caernarfonshire called for steps to be taken to devise means other than the holding of an annual show and an annual county meeting for keeping members together and furthering agricultural interests – the Caernarfonshire members suggested that this might be achieved by organizing demonstrations and lectures in connection with livestock from a breeding point of view in various parts of the country and by endeavouring to bring the county shows in Wales more directly under the influence and guidance of the national society than hitherto. Summing up their position, the Caernarfonshire members suggested 'that the Society should act as an Agricultural Society in the wider sense of the term not merely as a show-promoting body'. Cardiganshire, for its part, suggested that innovations, based upon the experience gained by County War Agricultural Executive Committees, might be introduced in the machinery section of the Royal Welsh Show. As a nice gesture to the value of tradition, the Cardiganshire county meeting submitted that, although tractors and lorries might, to some extent, displace heavy horses, there would always be a demand for Welsh cobs – especially for work on hill farms – and so it recommended to Council that adequate provision be made for the breed at future shows. One significant pointer to the future was contained in the recommendations from Denbighshire, Flintshire and Pembrokeshire that Young Farmers' Clubs be given a greater role in the future of the Society, as well as greater provision at the show for competitions sponsored by them.

Such proposals from county meetings – the Society's unofficial 'think tanks' – were always given a good airing at Council meetings. Much discussion took place in particular on Caernarfonshire's call for the Society to arrange demonstrations in the various Welsh counties. C. Bryner Jones, the new chairman, felt that the limited staff at the Society's disposal and the absence of a technical officer rendered it very difficult to implement such a proposal. At any rate, the fact that in future there would be an official staff responsible for agricultural education in each county led him to believe that they would be in a much better position than the Society to organize and carry out effective demonstrations on the lines suggested by the Caernarfonshire members. Even so, he recognized the need for the Royal Welsh, in common with other agricultural societies, to devote still more attention to the organization of agricultural educational demonstrations at future shows. In so far as the involvement of the Young Farmers' Clubs was concerned, Council decided that the forthcoming conference of the Young Farmers' Club movement at Aberystwyth in early April 1946 be invited to consider and report to the Council at their next meeting how they could best cooperate with the Royal Welsh Agricultural Society in connection with the Carmarthen show in 1947.

Much thought also went into deciding the appropriate time for holding large shows after the coming of peace. What impinged on the decisions of all the major agricultural societies in Britain was the ruling by the Labour government's minister of agriculture, Tom Williams, that shows could be held in 1946 only for a shorter duration. Having failed to reach an agreement on the morning of 20 November 1945 over whether or not to hold a two-day show in 1946, the decision was left by the Emergency Committee for the Council to decide later that day. While some of the committee had supported going ahead on the grounds that, despite all the difficulties in the way of advertising and printing, catering, transport, and of trade firms staging anything approaching a representative display, it was desirable to keep the Society before the public, others like Moses Griffith, Reuben Haigh and C. Bryner Jones felt that a mere two-day event would detract from the Society's prestige. Council members finally decided against holding a show on the grounds that it would be impossible to hold anything like a representative show in 1946.

During the war years it had been more or less a matter of carrying on the Society in skeleton form. The financial position had in fact improved over the period because a certain amount of income had accrued and expenditure had been modest. Some £500 had been received in membership subscriptions during the first year of the war, and although that level fell away thereafter, there was still an annual income from interest on investments. Overhead costs were negligible: a virtual suspension of its activities meant that the secretary was only re-employed full-time from 1 January 1944, followed later in the year by the re-engagement of Walter Williams as assistant secretary; and, as we have seen, thanks to Reuben Haigh's generosity, no expenditure was incurred through office hire. Only from late 1945 were offices once again retaken at Wrexham, this time for rooms in Regent Street. In October 1945, the Society's assets amounted to approximately £10,000, of which £1,763 was represented by cash at the Midland Bank, Aberystwyth. The post-war financial position was better indeed, confessed C. Bryner Jones, than he had anticipated.

1947–1963   Notwithstanding the huge public support for the Royal Welsh Shows at Carmarthen in 1947 and Swansea in 1949 (the 1948 event was cancelled owing to the reintroduction of petrol rationing), the Society's organizers were aware that the boom period, a form of public release after the wartime privation, would not last. Indeed, at the close of the 1940s they were in sombre mood about the Society's prospects, knowing that in future they would be called upon to meet much higher administrative expenses and costs in staging the show than in the past. Their gloomy prognostications were to be proved correct; the cost of staging the show annually rose from £15,271 in 1947 to £57,832 in 1962. Faced with this new situation they were acutely aware that their only *assured* annual income stemmed from interest on the Society's capital investments and from members' subscriptions. Although the three highly successful shows of 1947 (Carmarthen), 1949 (Swansea) and 1950 (Abergele) substantially boosted the reserves, its officials, with past experience in mind, appreciated that they could not rely on their shows to provide them with a ready source of income; sooner or later they would suffer losses upon some of them and a succession of financial flops as experienced in the shows of 1936, 1937 and 1939 would make extremely serious inroads into their capital and possibly even place the Society's future in jeopardy. Such a precarious position could only be rectified by a substantial increase in membership, which, standing at 2,188 at the close of 1950, was lamentable and unworthy of a national agricultural society. That total was even surpassed by English regional societies, the Shropshire and West Midland having a membership of 4,500 and the Yorkshire 8,528 members.

The forebodings of officials were to be realized often during the early and mid-1950s, for the Society underwent some lean and difficult years which cumulatively led to a serious erosion of its financial resources. There was a disturbing want of consistency in revenue from one year to the next. Although it was acclaimed as the best and most ambitious show since the war, the financial result of the Llandrindod Wells show staged at Llanelwedd in

1951 was disappointing for, despite the good weather, total income fell well below total outgoings, due largely to a fall in gate receipts compounded by an enormous increase in the charges for labour employed in erecting the showyard, cost and hire of materials and rising costs of printing. An outbreak of foot and mouth disease – which caused cancellation of the cloven-hoofed sections at Caernarfon in 1952 – aggravated by increasing recurrent costs, caused a deficit once again in accounts for that year. It was painfully clear that the financial results had changed markedly between 1947 and 1952. After profits of £12,680, £11,173 and £3,814 on the years' accounts for 1947, 1949 and 1950 respectively, the following two years witnessed losses of £2,435 and £2,552. So alarming was the situation at the end of 1952 in the face of inadequate revenue and mounting costs in every direction – showground erections, prize money, printing and stationery, advertising and bill posting, and police service – that the December Council, anxious to raise additional sources of income, agreed to raise the normal membership subscription in 1953 from a guinea, the level fixed in 1904, to £2 and to make increases in admission charges to the showground and trade stand charges. The tendency for expenditure to in-crease at a greater pace than income was to persist, so that even when the show was held in the populous district of Cardiff in 1953 the surplus on the year's accounts was only £728, thereby dashing hopes that the 1953 result would recover losses amounting to nearly £5,000 in 1951 and 1952. The upshot was that, in late 1953, the Council agreed that in future the aim should be to balance the budget each year wherever the show was held rather than rely upon a big surplus from a densely populated urban area. A serious setback to the Society's finances occurred in 1954 as a result of the rain-drenched, mud-deep Machynlleth show which affected the gate considerably. It meant that, despite reducing expenditure over the previous year by more than £4,000, the Society suffered on its year's working a deficit of £5,605, the worst in its entire history. Understandably viewing this as a matter of grave concern, Lieutenant Colonel G. E. FitzHugh, show director and chairman of the Finance and General Purposes Committee, urged upon Council the need for even closer investigation into the Society's expenditure in order to reduce it. Even the planned publication of a special jubilee brochure to celebrate the Society's first fifty years was abandoned. Fortunately, good weather, combined with a record local appeal fund of £11,325, ensured a highly successful show at Haverfordwest in 1955. The resultant handsome surplus on the year's accounts of £4,644 allowed the Society to recover some four-fifths of the loss incurred in 1954. For all the efforts at prudence, however, the Society still faced rising costs, and in 1955 expenditure reached a total of £40,874 against £38,093 in the previous year. As usual, the main increases were experienced in prize money, showground erection and show expenses.

The baneful influence of rising costs on the Society's fortunes was cruelly manifested in 1956. Although the show at Rhyl enjoyed record entries, perfect weather and a good total attendance of 64,000, the year's working led to a very substantial loss. Paradoxically it was the very size and success of the show, resulting as it necessarily did in greatly increased expense, which largely contributed to the year's deficit of £6,880, the largest the Society had ever experienced. In so far as show expenditure in 1956 was concerned, show-ground erections had reached a prohibitive figure – some £2,965 more than the previous year – while show expenses had also increased by £1,477. The crisis prompted a gloomy but candid statement by the chairman of the Council, Brigadier Sir Michael D. Venables-Llewelyn, regarding the dire financial straits in which the Society now found itself. Whereas at the end of the war the Society's capital had amounted to £10,666, a sum which a short post-war show boom period had increased to nearly £40,000, between 1950 and the close of 1956 it had steadily eroded to the dangerously low figure of £22,000. Essential in explaining this was the fact that the Society's 'ordinary' income, which represented income not directly connected with the show, always fell short of the expenditure necessary to fund the everyday running of the Society by more than £3,000 each year. If the show did not produce a profit to meet this deficit, capital had to be realized. Since 1950 (with the exception of 1955) the profit had not been sufficient for this, and, worse still, if the show itself resulted in financial loss then capital had to be withdrawn to meet both deficits. So fragile was the situation at the end of 1956 that an impending sense of doom pervaded the utterances of leading officers, a sinking realization that if improvement was not effected the Society could find its financial reserves exhausted within a very few years. The way out of this tightening financial strangulation was for expenditure to match income, by officials undertaking only what could be afforded and by avoiding subsidizing the Society's tastes and standards from the meagre and fast diminishing reserve fund. The Society simply had to live within its means. Costs were rising remorselessly – particularly showground erection expenses – and by over £8,000 in 1955 and 1956. Not only would expenditure have to be curtailed, but the pathetically low membership of 2,500 on 1 January 1957 would have to be significantly raised. It was estimated that not until a regular membership of at least 7,500 was reached could the Society's future be considered secure. Finally, from 1957 onwards it was stipulated that as far as the local fund was concerned the balance remaining after all expenditure undertaken by the local committee should be not less that £3,500, which would in due course be handed over to the Society so that the aforementioned deficit in its normal running costs could be met.

At the end of 1956 and in January 1957 serious deliberation was given to the size and type of shows which the Society could contemplate for the future. It was apparent that the size of the Rhyl show, for example, was too large for the expenses to be covered by the sort of attendance which could reasonably be expected in a less than densely populated Wales. Some reduction in show size would have to be accepted as a more or less permanent state of affairs. Indeed, once the difficult decision to go ahead with the Aberystwyth show was taken at the close of January 1957 at a time of petrol rationing, strenuous efforts were made drastically to prune show expenditure, thus permitting a balanced budget and holding out some prospect of a modest profit. Following a policy of strict adherence to the figures stated in the prepared budget, whereby economizing on expenditure to the tune of £10,000 was laid down, the Aberystwyth show, despite the sensational flooding of the ground on the final day, produced a very favourable financial result, and secured the Society a surplus in the annual accounts of £7,800, which meant that it was in balance as regards losses and profits since 1953. Mindful of the large Rhyl show, the consequences of heavy expenditure and the limited attendances that could be expected in Wales, it was aimed to keep the budgeted expenditure for the 1958 show at the level of the previous year and, in Lieutenant Colonel FitzHugh's view, the only physical way to limit the cost of the show was to restrict the size of the ground. That year, however, once again proved a disappointment, with the annual accounts showing a deficit of £3,700. Whereas the Society's income for the year fell substantially, some of this decrease arising from the unfavourable weather during the show, expenditure rose, largely as a result of an increase in show expenditure in showground erections, prize money and show expenses. The old Adam reasserting itself, this time committees had once more exceeded their budgeted allowances. For the future, the long-serving and supportive showground contractors, Messrs Woodhouse of Nottingham, were instructed to revert to the 1957 system wherein they had been able to keep a close watch on the day-to-day expenditure.

By now the large financial fluctuations that occurred from year to year were a regular feature and, true to form, 1959 was to prove a financially successful year. The Society's main source of income, the annual show – held at Margam over four days – proved a very successful event. Show receipts exceeded budget expectations by some £4,000, while expenditure was £1,100 above the anticipated sum. The Society's accounts for the year to 30 September 1959 revealed a surplus of £6,855. Thus, of the twelve annual shows held since 1947, some seven had produced a surplus and five a deficit, the last seven years from 1953 to 1959 producing surpluses and deficits in alternate years. Of importance in easing the transition to the permanent site at Llanelwedd in 1963 was the ending of the previous annual fluctuation cycle in the years after 1959. By 1961 it was happily apparent that the 'alternate year' sequence had been broken, the Society having enjoyed good surpluses in 1959 (at Margam, nearly £7,000), 1960 (at Welshpool, £11,000) and 1961 (at Gelli-aur, approximately £15,000). Indeed, this surplus of £15,000 made 1961 a record year financially, the result of a superb show held at Gelli-aur (Golden Grove), near Llandeilo. Ever the fly in the ointment, annual expenditure was still rising, however, and in 1961 it was £6,000 more than in 1960, with the biggest increase occurring in showground erections. It was fortunate, therefore, that the income for the year increased by almost £10,000. In 1962 the Society acquired its present site at Llanelwedd at a total cost of around £39,000. Fortunately, the handsome sum of £22,000 collected by the Builth Wells Permanent Site Committee, together with the 1961 surplus, allowed the Society to purchase the new site without dipping into its reserves. Doubtless frustrating was the loss of just over £7,000 for the financial year ending 30 September 1962; the total income showed a fall of £16,562 over that of 1961, mainly because of an appreciable drop in attendance at the Wrexham show compared with that of Llandeilo, and the total year's expenditure exceeded the 1961 figure by £4,731, largely due to increased labour costs and flower-show erections and costs incurred through the introduction of new competitions. The rising trend in annual total expenditure down to 1962 was worrying, and the Society's officials hoped that the trend would mercifully discontinue once the permanent site at Llanelwedd had been fully developed.

The migratory years 1947–62 witnessed many trials in the Society's financial affairs, especially 1956. It was noticeable that the best results were experienced when the shows were held in mid Wales, the south and the west – at Carmarthen, Swansea, Haverfordwest, Margam, Llandeilo and Welsh-pool. For all the difficulties and setbacks of these years, the honorary treasurer, J. E. Rees, looking back over the period of his stewardship since 1946, could point with satisfaction to the fact that the Society's meagre resources in 1945 had increased tenfold by 1962. From a reserve fund of merely £8,627 in 1947, net assets including the new permanent site had grown to more than £84,000 by 1962. Liquid assets, which excluded the Society's property at the Aberystwyth headquarters and at Llanelwedd, were in excess of £40,000.

Of course, there were underlying frailties which persisted until 1962 and beyond. Apart from the frustration and helplessness in the face of continued rising costs, during this last migratory phase no solution was found to the problem of sluggish membership, which so weakened the Society. As in the pre-war migratory years, one of the chief difficulties in the path of securing an increase in the permanent membership was that the show, alternating between north and south, was within easy reach of the majority of farmers every two years only and this deterred many of them from paying annual subscriptions. Despite the endeavours of the organizers to make it so, Wales's terrain and difficult transport problems hampered the Royal Welsh from being considered throughout Wales as a national show. Rather, too many persisted in thinking of it as a north Wales show one year and a south Wales show the next. By March 1960 the Council believed that the establishment of a permanent site was one important means of increasing membership of the Society. But the location of that site would have to be carefully chosen.

Since various other strategies for recruiting members had proved largely ineffective, attempts at boosting membership became ever more desperate in the face of mounting financial problems from the early 1950s. Apart from the long tradition of appeal to members through the pages of the annual *Journal* and letters of appeal sent by the chairman and the secretary, other approaches were tried: in 1954 various people were asked to act as agents on a commission basis; in 1955 sub-committees were formed in all Welsh counties; and appeals were made over the years through organizations such as the National Farmers' Union, Young Farmers' Clubs and the Milk Marketing Board. None of these produced any appreciable difference. Indeed, according to R. P. Thomas in a memorandum on membership prepared in 1962, the only method that had achieved any real success was the personal approach made to non-members by Josiah George of Roch, Pembrokeshire, who recruited no fewer that 102 members during 1951, and that made by T. H. Jones and Dan James in Carmarthenshire in 1957. In presenting his memorandum Thomas urged that since the permanent site had been purchased and membership stood at 3,649, it was imperative that this figure be raised to at least 10,000 if the Society were to survive on an economic basis, particularly as it would be without local funds in the future. In response to much discussion as to what should be done, Council decided in October 1962 that free parking facilities be extended to all members rather than a free issue of the annual *Journal*, that stricter discipline be applied for admission into the Council and Members' Pavilion, and that members be allowed reduced entry fees in every section of the show. For Moses Griffith, the explanation for the reluctance of Welsh farmers to become members lay in the Society's failure to address the pressing problems facing them, for example, their heavy livestock losses, and its failure to use the Welsh language as a medium for its activities.

We have noted earlier that from 1944 opinions had been aired regarding the future nature of the Society and its annual show. Acting in response to a call from Major John Francis of Carmarthen in January 1948 that the Council ought to have regard to the changed circumstances and to take steps which would enable the Society to fulfil its role and objectives, Council set up early in that year an Organisation Sub-Committee under the chairmanship of C. Bryner Jones. After two meetings, recommendations were forwarded to

Council and adopted in June 1948. Important steps were quickly taken towards implementing certain of these recommendations. For instance, summer 1948 saw the youthful appointments of Arthur George as secretary and John Wigley as assistant secretary, both of whom took office following the planned joint retirements of Captain Howson, secretary for twenty-one years, and Walter Williams, assistant secretary for thirty-eight years. Again, pleased at the return of its headquarters to its native Cardiganshire, Dr Alban Davies of Llan-non purchased new premises, Edleston House, Queen's Road, Aberystwyth, on behalf of the Society in January 1949 for £5,000, and generously offered to lend that sum free of interest to the Society for two years. Since he had been anxious that the headquarters should ultimately be transferred there, the late Lord Davies would have been delighted by this development. A further meeting of the Organisation Sub-Committee on 30 September 1949 made additional recommendations towards improving the organization of the Society's shows, such as providing better car parking arrangements, hiring paid stewards to work at the gates leading into the main ring, furnishing visitors with better information and signposting, and having announcers repeat the results of the awards.

Another important development emerged from a memorandum submitted by Lieutenant Colonel G. E. FitzHugh in April 1950 dealing with the weaknesses in the organization of the Council and its committees and making concrete suggestions in relation to their functions and procedure. The following June Council adopted certain principles of procedure, the main one of which stipulated that while crucial matters relating to the Society's administration and policy were to be reserved for the consideration of Council, such issues should not normally be initiated in Council but in the appropriate committee. Taken together, they were to mark an important step forward in streamlining administration, avoiding overlap, time-wasting, delay and confusion, and dealing more efficiently with the organization of the show.

By 1955 the number of standing committees had risen from six to eight, with the addition since 1950 of an Editorial Committee and a Horticulture and Honey Committee. Further changes to Council procedure and the constitution of committees occurred in 1956 in response to criticism in 1955 from D. O. Morgan that too little time was given to discussing matters arising out of committee reports in Council and too few elected members of Council were given an opportunity to serve on at least one committee, thereby debarring them from being sufficiently in touch with the Society's affairs. Certain recommendations were forwarded to Council by the Finance and General Purposes Committee, chaired by Lieutenant Colonel FitzHugh, to rectify its alleged shortcomings, notably that additional elected representation on Council should be granted to those counties whose membership of the

Society was strong, that elected Council representation should be bestowed on the border counties and those further afield, and that co-opted membership of Council should be restricted to twenty-five. Based on 1955 membership, the recommended new rules for representation on the Council would produce an approximate total membership of 110, of whom, and reversing the previous state of affairs, about two-thirds would be elected. Council approved these radical resolutions in March 1956. In so far as the number and composition of committees were concerned, the Finance Committee decided to recommend that the current eight standing committees of the Council should be continued – they comprised Finance and General Purposes, Stock Prizes and Judges Selection, the Honorary Director's, Protests, Machinery, Forestry, Editorial, and Horticulture and Honey – and that a Farm Produce Committee should be added. It was further recommended that where the majority of members of a committee were not members of the Council, its powers should be curtailed. Council similarly approved these recommendations in March 1956.

A further innovation in the election procedure for members of Council once again came in response to criticism. Accordingly, it was stipulated in October 1962 that a person was to be eligible for nomination for election to Council only after a qualifying period of three years as a member of the Society. By 1964, a fair measure of success had been achieved in ensuring that in electing committees every member had an opportunity to serve at least on one committee, though most Council members – reflecting the unavoidable, overwhelming and disproportionate representation of the livestock interest on the Council – preferred to serve on the Stock Prizes and Judges Selection Committee. Indeed, insufficient members of Council were willing, let alone anxious, to serve on some of the other committees which needed support. It is apparent, nevertheless, that in an attempt to achieve greater efficiency and to cater for wider public participation in the running of its affairs, the Society during these years had undertaken a radical reorganization of its Council and standing committees.

Over the years, absenteeism at Council and committees had been criticized by Society members. Part of the problem stemmed from the venue. With the exception of the Council meeting held on the showground during show week, the holding of other meetings at Shrewsbury – arguably the most convenient centre for most Welsh people – had long been unpopular with members from distant counties. It was in response to Carmarthenshire's recommendation in 1955 'that all Council meetings be held in Wales and that the venue be Aberystwyth' that Council in October of that year resolved that its spring and summer meetings would be held at Aberystwyth, while its autumn and annual ones would continue at Shrewsbury. This was still not the perfect solution, for Anglesey and Caernarfon Society members, through their 1956 county meeting, asked for meetings of Council to be held in north and south Wales alternately.

Another problem confronting the Society's leaders was the sparse attendance at county meetings in the 1950s. Merioneth had suggested in 1952 that a film show or a visiting speaker might be provided as an inducement to attend, though the idea was not followed up by the Society. Figures produced for 1957 revealed the seriousness of the problem: of the total membership of all county meetings of 2,214, only 151 members attended that year! So gloomy was the Council in the face of such apathy that, in 1958, it invited its Finance Committee to consider the principle of continuing with county meetings. That they were not wound up at this juncture was doubtless due to the universal desire of all counties, as expressed in early 1959, that county meetings be continued as at present. Nevertheless county meetings as they had been conducted since 1924 were to be superseded from 1961 onwards by county advisory committees in order to give counties more responsibility and to encourage their members throughout Wales to become more involved. As we shall discover in Chapter Nine, much was happening in the sphere of the Society's educational activities, and it was felt that the Society needed to promote a better relationship with the individual members so that these activities would become better known among the rank and file. Council members in each county were asked to form themselves into a nucleus of a 'county committee' so that periodical meetings could be held and demonstrations staged either with or without the cooperation of other bodies. These were an addition to the Society's constitution and, along with them, honorary county secretaries were appointed. At this time of huge challenge, it was hoped that the new county organizations would strengthen the Society as a whole. Furthermore, since the Society had now acquired a permanent site, there was the even more urgent need to build strong links with rural communities in all parts of Wales.

The relationship between the Society and the show's Local Executive Committee underwent some change in the period after the Second World War. Before 1939, as we have seen, the Society required the local committee to provide a ground and car parks, certain services, prize money for local classes, a bank (if desired) and a guarantee of up to £1,000 against loss on the Society's show account, which was then kept separately from the management account. Since 1947 no guarantee against loss was sought by the Society and, commencing in 1956, there were no local or district classes. Otherwise the requirements were unchanged. Yet, up to 1948 there was always a signed agreement between the local committee and the Society setting out the obligations. This afterwards lapsed, the matter resting upon goodwill without legal obligation. By the mid-1950s, as costs increased, the Society came to require an average surplus on the local fund of between £3,000 and £3,500. Only in 1955 was this achieved (£3,700). Experience had by this time shown that, in the absence of any legally enforceable agreement, great care would have to be taken, before accepting an invitation, to ensure that the necessary local support would be forthcoming and would be adequately organized. By 1956 it was felt that the Society's arrangements had not safeguarded its position sufficiently. With so much at stake, there was a need for a systematic approach which would ensure that, by the time an invitation had been received, support for the project could be accurately assessed and the key positions in organizing that support placed in reliable hands. With these concerns in mind, in September 1956 the Finance and General Purposes Committee recommended to Council the adoption of a specific procedure which amounted to the Society's making an unofficial approach within a certain district to determine the merits of a site with key local authority officials advising them. Upon receiving favourable signals for the go-ahead, Council should ensure that, in future, the local committee be asked to meet the cost of providing: the showground and parking spaces and the preparation and reinstatement thereof; the necessary services, such as electricity, water and access roads; and certain essential services enjoyed locally like police, fire and ambulance. Furthermore, it should provide a substantial contribution towards the cost of running the show. Finally, Council was recommended to adopt careful procedures in relation to privileges in return for donations to the local fund. Council endorsed all of these recommendations.

The post-war years down to 1962 witnessed growing provision for the use of the Welsh language in the Society's activities, although certain people, notably Moses Griffith, would have preferred the Society to have gone much further in this respect. Having appointed a Welsh speaker as secretary in 1948, in 1950 the Society decided to include 'Cymdeithas Amaethyddol Frenhinol Cymru' under the English title of the Society on all its letter headings and publications. Soon afterwards, in 1952, with due regard to the fact that the forthcoming show was being held in a Welsh-speaking district (Caernarfon), Council decided that commentaries in both English and Welsh – the first time that the mother tongue was used in the Society's history – should be given during the parades and Moses Griffith was appointed to deliver commentaries in both languages. Although the parade commentaries were in English at Cardiff the following year, bilingual ones were again given at Machynlleth in 1954, when Moses Griffith covered the horses parade, R. L. Jones the cattle parade, and E. J. Roberts the sheep parade. While no parade commentaries were held at Haverfordwest in 1955 because of the physical difficulties involved, at the Aberystwyth show in 1957 two commentators gave

25. Moses Griffith (left) at the Royal Welsh Show in 1959.

the main parade commentary, J. E. Nichols in English and Llywelyn Phillips in Welsh, and the same two acted as English and Welsh commentators respectively at the Bangor show the following year. It was Margam's turn to host the show in 1959, and here Professor Nichols acted as the chief main ring parade commentator, I. M. Yeomans covered the horses section and W. J. Constable the cattle one, while Llywelyn Phillips alternated in Welsh as required. Although more Welsh commentary was being given than in the pre-war era, getting the balance to suit everybody was understandably a delicate problem; on the one hand, the absence of sufficient Welsh commentary at Aberystwyth in 1957 was a source of grievance while, on the other, the Honorary Director's Committee meeting in September 1961 received criticisms of the parade commentary, in particular the sudden change from English into Welsh.

Discussion arose, too, in 1957 on the subject of translations of articles in the Society's *Journal* from English into Welsh and vice versa. However, no concessions were offered here, even to Moses Griffith's request that just a summary be provided in the alternative language. Influencing the Society's decision was the feeling that it was unfair to ask authors to include summarized translations in the other language.

At a meeting held on 4 March 1960, Moses Griffith reminded the Society that its relationship with the farming community could be fostered if it made greater use of the native language. In order to demonstrate the Society's palpable neglect, he pointed out that at the forthcoming 'Royal Welsh' Conference at Welsh-speaking Corwen in May the composition of the evening's Brains Trust should include a much stronger Welsh-speaking element. In the ensuing discussion R. W. Griffiths replied that, while the Welsh language was important, it was not of primary importance, and he posed the question whether Welsh people could achieve a higher standard of living by relying on the language. As far as he was concerned, there should be a wider vision. Lieutenant Colonel Beaumont, however, called for more attention to be given the Welsh language.

Crucial, of course, to policy-making and the planning and running of the Society's affairs in these years were its staff and key officers. The expansion of activities in the period following 1947 led to greater burdens on the staff, even though the complement had increased from five in 1949 to eleven by 1959. Senior staff in this period of migratory shows spent a significant amount of their time 'on the road', so much so in the case of its secretary, Arthur George (who between July 1959 and June 1960, for instance, travelled on 165 days), that in a bid to save time the Society agreed in 1956 to purchase a unit costing £22 for adapting his tape-recording machine to be used from the battery of the car! In terms of human resources, 6½ persons were engaged in Royal Welsh work between July 1959 and June 1960, four persons on Welsh Pony and Cob Society administration and one person for half the year on Welsh Mountain Sheep Society and Welsh Halfbred Sheep Breeders' Association work.

Arthur George, a Welsh-speaker, was brought up on a farm in north Pembrokeshire and came to the Royal Welsh after a spell as liaison officer for Wales for the National Federation of Young Farmers' Clubs. He succeeded Captain Howson as secretary in September 1948. Along with the assistant secretary, John Wigley, who took office at the same time after having already served the Society for two years looking after the breed societies, he brought youth and energy to the Society. Like so many others who had held office within the Society – for instance Walter Williams, whose thirty-eight years of service as assistant secretary before he retired in 1948 absorbed practically the whole of his time and thought – Arthur George during his working life with the Society until 1973 was a self-styled 'workaholic'. Although, unlike Captain Howson, he was not a gifted and prolific writer, he was an excellent administrator and organizer and, equally vital, able to get on with people. Thus, much of his effectiveness lay in the constructive, cordial relationships he established with his chairmen, notably in these early years with C. Bryner Jones and Lieutenant Colonel G. E. FitzHugh. As well as his huge contribution in

26. Arthur George (second from right), secretary from 1948, with Field Marshal Mont-gomery at the Swansea show in 1949.

facilitating the transition from migratory show to permanent site, his time in office was marked by his evaluation of all the Society's activities from an educational point of view. Possessed of a nice sense of self-deprecation, Arthur George relates in his private papers one amusing slip early in his career at the Royal Welsh. As part of the preparation for the show to be held at Abergele in 1950 he made reservations at a certain hotel 'on the front' at Rhyl which, in the light of its good reputation for food and comforts, he deemed to be suitable accommodation for the main ring stewards, including Major Jack Rees, Percy Thomas and Ben W. Rees among others of 'the team'. George continues: 'Close acquaintances of these gentlemen can imagine the "uncomplimentary abuse" I had to suffer when they discovered on arrival that the establishment was a temperance hotel. I was never allowed to forget my error.'

Dedicated service, too, was given by the honorary treasurer, J. E. Rees, manager of the Midland Bank, Aberystwyth, who between 1946 and 1961 carried on the good work that the bank's former manager, R. H. Thomas, had done before him since 1927. Such, too, was the latter's devotion to the Society's welfare that, despite being a very sick man, he had insisted on accompanying J. E. Rees to Carmarthen in 1947 to guide him at his first show. Thomas's illness in fact meant that he returned home on the afternoon of the first day, and he died soon afterwards in November. J. E. Rees was honorary treasurer during some very bleak financial years, and he must he have felt a sense of satisfaction upon stepping down in 1961 at a time when the Society had enjoyed three good surpluses in succession. The quiet dedication of this man is reflected in his preference during show week for lodging at a private house where solitude allowed him to study figures, turnstile readings, schedules and the like in time for delivery of statistical details to officials and staff the following day. He would doubtless have been aware of the heavy drinking and high jinks at the hotels among the show stewards. Graham Rees, long-standing steward for Welsh cobs at the show, recalls that during this period of the migratory show 'apple pie' beds were part of the frolics and that Captain Bill Williams, one of the stewards and a retired army officer who had served in India, would frequently take down the curtains from the hotel lounge, proceed to dress himself up, complete with turban, and perform the Indian rope trick!

Of immense importance in shaping the direction of the Society during the entire first half century of its history was C. Bryner Jones (knighted in 1947), who, after holding the professorship of Agriculture at Aberystwyth from 1907, became agricultural commissioner for Wales in 1912. Thereafter, until his retirement from the post of Welsh secretary to the Ministry of Agriculture in 1944, he occupied a unique position at the forefront of Welsh agricultural development. In Sir George Stapledon's words, he was 'the father of agricultural research and education in the Principality'. A man of great integrity, personal dignity and unfailing courtesy, his first dealings with the Society came at its inception when, as a lecturer at Newcastle, he was invited to judge the machinery section competitions at the first show. Understandably, his move to Aberystwyth University College in 1907 strengthened the ties; joining the Society in that year, he was elected a member of the Council in 1908, a body which he served faithfully to the end of his days in 1954. Owing to the illness of Lewes Loveden Pryse, in 1909 he was appointed honorary director of the show and secretary (*pro tem*) of the Society. Mention has been made also of the principal role he played in persuading David Davies of Llandinam in 1921 to revive the Society following its lapse since 1914. In the unavoidable absence of chairman David Davies from a number of meetings, he filled the breach and in 1931 he was fittingly appointed senior vice-chairman of the Council. There was by now an air of inevitability about his succeeding Lord Davies as chairman upon the latter's death in 1944, and he continued to serve that important, demanding office with distinction until 1953 when – a

27. Sir C. Bryner Jones, president, at the Machynlleth show, 1954.

28. Sir Michael D. Venables-Llewelyn, chairman of the Council from 1954 to 1969.

crowning honour to an illustrious career – he was elected president of the Society in its jubilee year, 1954. Sadly he died during that year. The fact that he was absent from Council only once and that he missed only one annual show is shining testimony of his devotion. For him, the show was important both as a shop window of the agricultural industry in Wales and as an educator. In particular, he was to concentrate upon the machinery section, seeing it as a vital means of improving farm implements and machinery best adapted for the distinctive needs of Welsh farmers and farm workers. If, perhaps, it was for his work outside the Council chamber that the Society was most indebted to him, he was nevertheless a model chairman, in which office he was well served by his clarity of thought, critical appreciation and sound judgement. Arthur George, at his side throughout as secretary, was to recall that 'not only was Sir Bryner perfect on procedures, he was equally brilliant in their execution following the decisions taken, be they by way of report or letter or by personal contact'.

Two more names came to the fore in the 1950s with the passing of old stalwarts of the Society. Sir Bryner Jones's successor as chairman of the Council in 1954 was Brigadier Sir Michael D. Venables-Llewelyn of Llysdinam Hall, Newbridge, whose father, Colonel Sir Charles Venables-Llewelyn, himself president of the Society in 1932, had presented his young son with a life membership of the Society as a Christmas present. Sir Michael had proved his mettle in 1951 as both president of the Society and chairman of the Llandrindod Wells Local Committee when the show was held at Llanelwedd. He was to hold office for fifteen years down to 1969, during which he was, in Lieutenant Colonel FitzHugh's words, an 'ideal' chairman. In particular, Sir Michael was expert at drawing up a programme and itinerary on the occasion of royal visits to the showground. Readers will recall too the unfailing service given to the Society by Reuben Haigh as honorary director over a period of twenty-three years. Upon his resignation in 1950 due to ill health, he was succeeded as honorary director by Lieutenant Colonel G. E. FitzHugh of Plas Power, Wrexham, who played an inestimable role in the Society's affairs both before and after the move to Llanelwedd in 1963. It must suffice here to record once again that in his capacity as chairman of both the Finance and General Purposes Committee and the Honorary Director's Committee in the 1950s and down to 1962 he was the key figure behind the reorganization of the Council and its committees and he was likewise to the fore in steering the Society through the choppy waters leading to the permanent site. We shall review in Chapter Ten this great step forward.

CHAPTER EIGHT

The Show, 1947–1962
EXPANSION AND INNOVATION

Attracted by the knowledge that the show was not confined to farmers only and that there was something in it to appeal to all sections of the community, the public, who increasingly arrived in private motor cars, attended in larger numbers after the war. Despite excellent cooperation from the various county police forces, a long traffic queue became part of the Royal Welsh scene during these years. Unsuspecting visitors, for instance, were caught up in the four-to-five-mile crawls to the Haverfordwest show in 1955 and Gelli-aur in 1961. Three-day show crowds ranged between 50,000 and 60,000 at the lower level (falling to 45,375 at rain-drenched Machynlleth in 1954) and 80,000–90,000 in the upper range. Gelli-aur, Llandeilo, in 1961 attracted the highest three-day gate of just over 90,000 although, amazingly given the freak flood on its third day, Aberystwyth's total attendance in 1957 was just over 83,000. Doubtless Princess Alexandra's visit to that particular show greatly boosted attendance. Swansea's four-day show in 1949 drew a crowd of just over 102,000. Excluding figures for the four-day events staged at Swansea (1949) and Margam (1959), the average attendance of the three-day shows between 1947 and 1962 was around 65,000. As such, gates were considerably larger than in pre-war days, when crowds averaged some 35,000 and did not generally exceed the 40,000 mark, apart from Carmarthen's record attendance of nearly 53,000 in 1925.

The big crowds, especially on the second day when often more than 30,000 people thronged the avenues, meant that showgrounds, despite the increase in their size to around fifty to sixty acres, were frequently cramped and overcrowded, and services and facilities overstretched. We shall see that one of the reasons for deciding upon a permanent site was the response to growing criticism from stock exhibitors and show visitors alike of poor facilities, particularly those relating to hygiene and the provision of food. Traffic arrangements at the 1961 Gelli-aur show, particularly the congestion at the entrances, were criticized by members of Council, who pointed out that this inconvenience was developing into a serious charge against the Society and was a problem that should be given priority. Naturally, conditions worsened at times of rain and flood. The state of the ground at Machynlleth in 1954 was so poor that 'Landsman' of the *Western Mail* predicted that it would

29. Machynlleth show, 1954: Valerie Thomas and Norton Jones braved the mud to examine the bogged-down vehicles.

be recorded in the annals of the Society as a 'gumboot show' while, unfeelingly, the agricultural correspondent of the *Daily Express* dubbed it the Royal 'Squelch' Show. More positively, *Y Cymro* carried the heading: 'Great show in spite of rain and mud'. The preceding ten weeks of heavy continuous rain had converted the ground into a quagmire and, in spite of all the valiant efforts at laying down railway sleepers and scattering straw and gravel, the atrocious conditions underfoot remained throughout the three days. Four caterpillar tractors were busily employed pulling out vehicles stuck in the sea of mud. Indeed, the final day, 23 July, was for 'Landsman' the worst day he had experienced in a long history of attendance at the Royal Welsh. The heavy rain throughout the day had reduced the avenues to a morass of mud, which bogged down all transport and cut off the supply of foodstuffs for the different canteens. All the more was this disappointment given that 1954 was a special year for the society. Sir C. Bryner Jones had remarked earlier in the year: 'The Royal Welsh, as a Society, this year celebrates its fiftieth anniversary, and every effort will be made to make the Show in Machynlleth one that is worthy of this occasion and a Show that is worthy of Wales.' Part of the problem in churning up showgrounds in periods of rainy weather stemmed from the use of heavy

30. Bangor show, 1958, showing poor conditions underfoot.

traffic vehicles, as was plainly the case at Bangor in 1958 where the ground was in a deplorable state over the first two days. A wet period from June onwards had rendered the site sodden, a condition that was converted into sticky mud by heavy lorry traffic before the show opened on the first day.

Although the problem would only truly be redressed by the provision of hard road surfaces, some definite improvement in conditions underfoot for the public came from 1959 onwards with the introduction of separate pedestrian and vehicular avenues. At Margam in 1959, for the first time ever, lorry traffic before and during the show was confined to the back lanes, so that the frontage roads were always free to roaming visitors. So successful was the experiment that it was retained at Welshpool the following year. In fact, the Welshpool show contained several 'new departures' in layout and accommodation designed to promote the comfort of the visiting public.

Increasing numbers of traders' stands provide a good indication of the show's growth and popularity during these years. Whereas there had been an average of 151 per show at the six shows held immediately down to and including 1939, that average increased to 260 for the years 1947 to 1962. There was a growth, too, over these post-war years themselves: the average number of traders' stands during the five shows staged between 1947 and 1952 of 210 rose to one of 306 over the five held between 1958 and 1962. It was here in the various trade displays that the astounding increase in farm mechanization was fully reflected. By the mid-1950s the horse had virtually disappeared and almost every farmer possessed at least one tractor together with other power-operated machinery. So anxious was the Society to concur with the wishes of this vocal group of trade-stand holders that from 1961 onwards the traditional days of the show of Wednesday, Thursday and Friday were altered to Tuesday, Wednesday and Thursday in order to enable them to leave in time for the next weekend or to reach the Royal Lancashire Show. As noted above, bigger crowds and mounting numbers of trade exhibits meant larger sites. Whereas the area of the showground at Welshpool in 1923 was a mere twenty-three acres, its site in 1960 extended to no fewer than sixty-two acres.

Despite the financial problems besetting the Society in the 1950s, the show manifestly expanded in scale and variety. This, in turn, called for ever more vigilant management. Certainly the setting up of an Honorary Director's Committee in 1950, under the able and dextrous chairmanship of Lieutenant Colonel FitzHugh, to deal with details of show organization brought clearer and closer oversight. From the mid-1950s, in particular, steps were taken to effect an improvement in showground organization. Thus, greater co-ordination came with the adoption of the practice from Machynlleth (1954) onwards of the Society's secretary also acting as secretary to the Showground Committee. In December 1955 Council also approved recommendations from the Honorary Director's Committee concerning the need for certain adjustments in the show and stewarding organization to make for better efficiency and complete control over all sections of the show. Henceforth, seven assistant honorary directors were to be appointed for shows instead of the three hitherto responsible for horses; cattle, sheep and pigs; and showground respectively. Each would take care of one of the following sections: horses; cattle, sheep and pigs; showyard (preparation and maintenance); showyard (administration); traders; competitions; and horticulture and honey, and would be responsible to the honorary show director. The next group in the chain of responsibility were to be thirty-five sectional senior stewards, who would be answerable to their respective assistant honorary directors. In addition there were to be permanent assistant stewards. These were judged to be such key posts that it was important for the smooth running of the show that they should be entrusted to the same person each year. They likewise were to answer to their respective sectional stewards. Although not meeting every Council member's approval, on the grounds that it had always been considered an honour in itself to steward at the Royal Welsh Show, it was decided that in the interests of retaining valuable men and also of introducing young blood into the running of the organization these permanent assistant stewards should be given help towards out-of-pocket expenses, since they were required to be on the showground for practically a whole week. Innovations, too, came in the mid-1950s as a means of improving the publicity given to future shows; information was to be in both English and Welsh and the latter was to be used in the press for localities where Welsh was the everyday language.

From its earliest days the show had responded to new developments in Welsh agriculture in a determined bid to move with the times. This capacity to adjust and to innovate was fully demonstrated in the years after 1947. An innovation at the 1951 show took the form of competitions for implements specially designed to work under conditions prevailing in the hill districts of Wales. Calls at this time upon the farmer to increase production from his land were frustrated within much of the Welsh countryside by the scarcity of implements suitable for such conditions, and the Society had begun to consider this problem from 1948 with the setting up of a special sub-committee to look into the steps necessary to encourage the manufacture of machines and implements best suited to Welsh upland areas. Out of this initiative emerged the Machinery Committee in 1951. Prizes were accordingly offered in 1951 in an open competition for a new, or substantial adaptation of an existing, implement or machine for use in connection with land drainage or reclamation and the cultivation of marginal and hill land. The winner of this open competition was the Brockhouse Engineering Co. of Southport for a lightweight tractor deemed to be potentially valuable for working small marginal or hill farms. A second competition for the best modification or adaptation of an existing implement or machine, confined to persons engaged in or closely connected with agriculture in Wales, attracted only one competitor, and although it was cancelled in 1952 and 1953 because of the paucity of entries, this competition for amateur home-grown inventors was revived in the 1954 show. At the 1956 show – where the pre-war Medal Award Scheme for new machines and implements staged by manufacturers on the trade stands at the show was resuscitated – a silver medal was awarded to a trade-stand holder who exhibited a new or modified machine or implement likely to make the most useful contribution to meet Welsh upland conditions. The seven entries for this Silver Medal Competition at Rhyl marked an encouraging start, and the two winners, the Salopian Trailer and the Wolseley 'Swipe', quickly proved beneficial to Welsh hill farmers, according to Colonel J. J. Davis, the judge of the competition. It was the turn of Massey-Harris-Ferguson with the Ferguson 35 Tractor to win the silver medal at the 1957 show. This year, too, saw the awarding for the first time of the D. Alban Davies Trophy, a special award to the entry most suited for upland conditions.

Its winner was Messrs R. A. Lister Co., Ltd, for the Lister Sheep Shearing Table. The activities of the Machinery Committee since 1951 in promoting machinery specifically suited to Welsh upland farming conditions were seen to be bearing fruit with the record fifteen entries at Bangor show in 1958 for the Silver Medal Awards. This represented an appreciation by the manufacturers of the services rendered by the Society to improve the livelihood of hill farmers in Wales. The winner of the D. Alban Davies Trophy was Messrs Massey-Harris-Ferguson of Coventry for the 725 Forager. At the same time, we must be cautious about claiming too much in the way of achievement; as late as 1961 a Society official recognized that 'the Show has not yet succeeded in getting machines specially designed for Welsh upland conditions, yet we [the Society] are paving the way towards that ideal in several directions'. It was recognized that the scheme was failing in its present form to offer scope for the inventiveness of the farmer.

If these machinery and implement competitions were arguably the major innovation at shows in the years following the war, there were also other new departures. Visitors to Rhyl in 1956 witnessed for the first time a display by the International Wool Secretariat in the form of a fashion parade under the title 'From Fleece to Fashion' by mannequins from London, who paraded daily and gave a preview of autumn wool fashions, all the garments shown being available in shops throughout Wales. Such was its success that the International Wool Secretariat once again staged an exhibition at Aberystwyth in 1957, including mannequin parades, to demonstrate the varied use of wool fabrics.

Staged next to the International Wool Secretariat, sheep-shearing competitions were introduced at the 1959 show at Margam, the Society having discussed the implementation of this new venture since 1957 with the New Zealand technique, currently demonstrated at the Royal Welsh Show by Godfrey Bowen, in mind. The 1959 competitions contained two classes, the first one 'open' and the second confined to members of the Young Farmers' Clubs. Following the 1959 competitions and demonstrations, improvements were made by installing a raised platform for the competitors and seating accommodation for the spectators. From 1960, also, the competitions would be referred to as the All Wales Championship Sheep-Shearing Competitions. The leading top shearer at the 1960 show was Isa Lloyd, a 24-year-old Englishman from Herefordshire, who had moved to Newchurch, near Kington in Radnorshire, in January of that year, and who had first entered sheep-shearing competitions at the age of fifteen. Although placed just fourth in that 1960 competition, his younger brother Sam was to share the prize for the champion sheep-shearer of Wales competition at Gelli-aur in 1961 with Eifion Evans of Llansannan, Denbighshire. Thirty-year-old Eifion sheared on a contract basis for north Walian farmers with two New Zealand shearers and had spent seven months in New Zealand two years previously learning the Godfrey Bowen – the world champion shearer – style. He was forced into second place at Wrexham the following year by Sam Lloyd, who had recently come to use the Godfrey Bowen technique after reading his book and being taught a lot about it, too, by his rival Eifion.

Several innovations were to feature at Gelli-aur in 1961, particularly the special efforts made to stage carcass competitions for lambs in conjunction with the National Sheep Breeders' Association. This new departure arose out of discussions at the Stock Prizes and Judges Selection Committee in November 1960 of the memorandum prepared by Austin Jenkins on the future pattern of livestock classes at the Royal Welsh Show. This document pointed to widespread discrepancies between the appraisal of live animals and their carcass characteristics in terms of meat and also between appearance and milk production. The committee agreed that while the appraisal of appearance should not of necessity be abandoned, it should nevertheless be placed in proper perspective in relation to performance and production characteristics. If the scale of the exhibit at Gelli-aur was modest and the facilities meagre, it nevertheless indicated future trends and marked the beginning of an important new phase in livestock competitions at the Royal Welsh as well as at other major shows. The encouraging reports of the 1961 competition led the Society to stage another in 1962 at Wrexham, where the first and second prizes were won by the brothers George and David Hughes of Rhyl, who were also winners of a new competition that year for Welsh half-bred sheep. Similarly, a pig carcass competition was organized at the 1962 show. The 1961 show at Gelli-aur, too, witnessed a new wool competition based on the same lines as that conducted at the Royal Agricultural Society of England Show. This change in the wool competition was popular as the old competition had attracted few entries.

New competitions for livestock breeds were introduced at shows over these years. Ayrshire cattle were exhibited for the first time at the Carmarthen show in 1947. Jerseys first appeared in open classes at a Royal Welsh Show at Abergele in 1950 and Red Polls were introduced at the Llandrindod Wells show (staged at Llanelwedd) the following year. At the 1956 show in Rhyl, the Landrace pig competition attracted more than forty entries. The Aberystwyth show the year following saw classes introduced for the Border Leicester breed of sheep. Likewise, the Gelli-aur show in 1961 included two new sheep classes, namely, Llanwenog and Beulah Speckled Face, while, as we have seen, a new competition at the Wrexham show in 1962 was instituted for Welsh half-bred sheep.

A significant change of practice came in 1954 with the decision to exclude local or district classes from future show schedules. One of the points made at an ad hoc committee in September 1954 was that the Royal Welsh Show, as one of the accepted 'national' events, should not lower its standards by including district classes, since the standard of entries in the latter generally fell below that of a county show. The following month, Council resolved that district classes would be excluded from 1956 onwards. This decision was also implemented at the 1961 show at Gelli-aur, despite the 'strong feeling' locally in Carmarthenshire that provision should be made for local classes. Ironically, after the decision to abandon the local classes had been taken, the normally uneventful course followed in these classes was sensationally overturned at the Haverfordwest show in 1955 when Wychwood Esprit, an eight-year-old Jersey cow belonging to L. A. Howell of Llechryd, defeated Cowin Rose, the previous day's winner in the open Jersey classes.

Changes and innovations also occurred in the parade commentaries at the show after the Second World War. It has already been noted that the Welsh language was used for the first time in 1952, when Llywelyn Phillips began to serve as the regular Welsh-speaking commentator at the migratory shows in the years ahead. A novel feature of the parade commentary in 1953 at Cardiff – though it did not become regular practice thereafter – was that it was split up into short talks by experts on each particular breed, a veritable seminar in the park. Clearly, the didactic potential of the commentary was being recognized and developed. This led Professor J. E. Nichols of Aberystwyth, a prominent member of the Council, to observe in 1959 that the Royal Welsh parade, plus commentary, now meant something quite different from any similar parade at other large shows. His advice to his colleagues in the Society was that the parade commentary should be purely objective and educational, avoiding both mention of individual breeders and stress on any specific breed. Professor Nichols was to return to this point in late 1961 when he claimed that the Royal Welsh had gone a great deal further than any other Society in creating an interesting main parade. One memorable performance occurred at Llanelwedd in 1951, where, according to 'Landsman' in the *Western Mail*, the 'outstanding achievement' of the second day was the 'brilliant' commentary on the grand parade of winning animals made at a moment's notice by Alan Turnbull.

Another new departure occurred at the 1960 show in Welshpool when the available space devoted to educational exhibits was extended. Eight acres were now taken up not only to give service to the farmer but also to demonstrate some of the new highlights of progress both in the scientific as well as the practical fields of farming. The trend was continued at Gelli-aur in 1961 where more attention than hitherto was paid to practical demonstrations and educational exhibits. As part of this the National Institute of Agricultural

Engineering staged an exhibit in conjunction with the Royal Welsh Agricultural Society to inform the farmer of the facilities they offered, particularly in operations encountered under Welsh farming conditions.

Despite the innovations and new look of the post-war shows, many of the familiar and popular features of those of earlier years continued to be big attractions. If by the early 1960s livestock were fighting to hold their own at the show before an invasion of massed machinery, this section, with its six sub-sections of horses, cattle, sheep, pigs, dogs (when asked for) and fur and feather, remained at the heart of the show (see Appendix 2). The rapid mechanization of Welsh farms following the Second World War, as seen in the near threefold increase in the number of tractors between 1945 and 1954, had witnessed the virtual demise of the horse on Welsh farms. Although there was some – though not a sharp – fall in the average number of entries in the Shires Section in the post-war Royal Welsh Shows, competitions were still a great focus of interest and doubtless evoked considerable nostalgia. Inspecting sixty-one great shire horses entered at Rhyl in 1956, all parading proudly

31. Wishful Select, Supreme Champion Shire Horse at Wrexham in 1962, owned by Trevor Thomas, Cwm Mawr Farm, Three Crosses, Swansea.

notwithstanding the rows of gleaming tractors and farm implements lining Machinery Avenue, Llewellyn Joseph of Porthcawl claimed that they were 'a sheer joy to judge and the standby of a day when the tractor runs out of oil'. To those who argued that there was always a place for one horse on the farm, the average forty-three shires entered per show over the years 1947–62 (compared with an average of fifty-five at earlier shows between 1922 and 1939) would have been a gratifying sight. Indeed, never to be forgotten by anyone who saw them at Rhyl in 1956 was that great class of shire mares and foals, with Hillmoor Sunset belonging to Messrs Richardson of Frogmore Farm, Moreton-in-Marsh, taking first prize, and the strong class of geldings led by the subsequent champion in the shire classes, Messrs Whewells'

Heaton Gay Lad. Here surely was conclusive proof that the modern shire, active and powerful, was not a spent force. In similar vein, even though horse-shoeing was a dying craft in the 1950s, competitions at the Royal Welsh – even if want of entries saw their cancellation at Cardiff in 1953 – attracted competitors of a high standard well into that decade. Carmarthenshire smiths won the cup at the three successive shows after the war and particularly impressive was Trevor Lloyd of Dolgarreg Forge, Llanwrda, who, having won the championship of Great Britain at the recent Royal Show in Oxford, swept the prize list at Abergele in 1950.

The Royal Welsh would not at any time have been a show to speak of without its classes for the native ponies and cobs. They have always been the highlight of every Royal Welsh Show, but even if their numbers at the post-war shows were to increase significantly compared with those of pre-war days there was still a disappointing turnout of Welsh cobs, reflecting their displacement under the remorseless encroachment of affordable little grey Ferguson tractors on to Welsh farms. There were a mere twenty-six entries of in-hand Welsh cobs at the 1947 show and even at Aberystwyth in 1957, set in the midst of true cob country, I. Osborne Jones was to judge a line-up of only twenty-nine. By virtue of the traditional cob-breeding families, under the prompting of the Welsh Pony and Cob Society, consciously working in the 1950s to revitalize the breed its great revival would happily come about from the 1960s. Despite the limited entries between 1947 and 1962, Royal Welsh competitions were keen and crowd-drawers. An important new trophy was presented to the Society in 1950, namely the Tom and Sprightly Perpetual Challenge Cup, by the hugely gifted horseman Tom Jones Evans, which was awarded annually for the best exhibit in the Welsh Cob, Welsh Pony, Welsh Mountain Pony and Hackney classes shown in hand or harness, and, uniquely, to be judged by popular applause. It was presented by him to commemorate both his association with the Society since its inception in 1904 and his favourite horse, Sprightly. This marvellous Welsh Mountain pony stallion had been depicted as an 'Idol of Idols' by William Evans in 1961 and was, in the latter's view, only to be matched in another section at a later date by the famous Welsh cob stallion Pentre Eiddwen Comet, bred by J. O. Davies of Llanddewi Brefi, and owned from 1950 onwards by John Hughes of Llanrhystud, near Aberystwyth. Such was the quality of Comet that he won the Tom and Sprightly Cup no fewer than five times, four of them at successive Royal Welsh Shows. Comet became a star at Llanelwedd in 1951, home of the Llandrindod Wells show, where he enthralled the crowds by winning not only the coveted George Prince of Wales Perpetual Challenge Cup but also, to tumultuous applause, the Tom and Sprightly Cup on its first offering. The astounding successive wins of this trophy began at Rhyl in 1956, continuing through to Margam in 1959. This run of triumphs was made all the more impressive by the fact that the fiery Comet had to prove himself in the face of worthy opposition from magnificent rivals such as Meiarth Welsh Maid, Mathrafal Eiddwen, Mathrafal Brenin, Myrtle Welsh Flyer, Brenin Gwalia and Llwynog-y-Garth. The last-mentioned cob stallion, for instance,

32. Welsh Cob stallion, Pentre Eiddwen Comet.

33. Welsh Cob mare, Parc Lady.

won the 'applause' cup at Welshpool in 1960 and, in the previous twelve years under the ownership of A. D. Thomas of the Grange Stud Farm, Neath, had won well over 500 first prizes and appeared at most of the major horse shows in Great Britain. Royal Welsh crowds from the late 1950s into the beginning of the 1960s were also enthralled by a lovely old Welsh type of cob brood mare, Parc Lady, owned by Daniel Morgan of Coed Parc, Lampeter – in the heart of cob country – whose prowess was such that she won the George Prince of Wales Perpetual Challenge Cup for the best cob of the old Welsh type at successive shows from 1958 to 1961. When he was seventy-three years old in 1961, Morgan recalled that before the Second World War he had used his horses for general farm work, but since the large-scale introduction of the tractor he had developed them for riding.

Prominent as an exhibitor of Welsh ponies at the post-war shows was Margaret Brodrick of Abergele who, in 1924, with her friend and business partner, John Jones, started the Coed Coch Pony Stud. Concentration on the best strains and a measure of line breeding produced a fixed type of striking beauty and superior quality, so that in later years Coed Coch ponies carried off numerous prizes in the show ring and became world famous as a result of her pioneering efforts at winning overseas markets. Her celebrated grey pony, Coed Coch Siaradus, won the Second Sprightly Challenge Cup for the best Welsh Mountain pony outright by virtue of three consecutive wins at the

Royal Welsh Show between 1950 and 1952. In 1951 the Second Sprightly Challenge Cup, the Kilvrough Challenge Cup and the breed championship all went to this magnificent pony, with Miss Brodrick's grey four-year-old stallion Coed Coch Madog in reserve! At the 1955 show at Haverfordwest, once again Miss Brodrick took five firsts, two challenge trophies and two medals with Coed Coch Siaradus and Coed Coch Siwgran. Two years later at Aberystwyth her by now ten-year-old stallion, Coed Coch Madog, whose show successes to date must have run well into three figures, again won the Society's medal for the best stallion or colt, though the supreme championship that year went to Emrys Griffiths's Revel Spring Song, an attractive cream pony bred by him at Talgarth.

34. Welsh Mountain pony mare, Coed Coch Siaradus.

Pedigree mountain ponies exhibited at the various Royal Welsh Shows in these post-war years were put up for sale and purchased by overseas buyers. At Abergele in 1950 several American buyers bought these ponies, nearly all of them prizewinners and most of them owned by Miss Brodrick. Later, in 1958, forty Welsh ponies from the rain-swept moorlands of north Wales, and exhibited at the show in soggy Bangor, were happily destined to spend the rest of their days in the more sunny climes of California after being purchased by Mrs Bonnie Parke of Twin Falls, Idaho, for £7,000. However, not so Shan Cwilt, the champion pony and winner of the Queen's Cup presented that year for the first time, for her owner, Mrs W. E. Morgan of Wellfield, Carmarthen, refused an offer of £700 for her! Catching the eye of foreign buyers at the Royal Welsh, the reputation of these ponies grew, so that more and more Welsh farmers in the 1950s found a very profitable sideline in breeding Welsh Mountain ponies for an increasingly demanding overseas market, especially in the United States and Canada. By 1961 it was no exaggeration to claim that the Welsh Mountain pony had found a second home in the United States.

Whereas by the end of the 1950s heavy horses and hunters were not required to remain for the whole period of the show, which resulted in considerable economies to the Society, cobs, ponies and other livestock exhibits remained for the entire three days. But the number of entries in many breeds of cattle by the end of the 1950s compared unfavourably with those of a good county show and, according to Lieutenant Colonel FitzHugh, one reason for this was alleged to be the long absence from home. It appeared that the expense and inconvenience of exhibiting meant that the tendency was to bring forward only those animals likely to win. From the 1947 show at Carmarthen, too, no entries were accepted from unattested herds, and, though the Society had no alternative but to make this decision, it had the effect of encouraging exhibitors to show only the best animals. Another factor, as in pre-war days, in reducing numbers of show entries was the reluctance of farmers in north Wales to travel and exhibit at shows in the south and vice versa. This was clearly in evidence at the 1949 show at Swansea where the great majority of entries in the Welsh Black classes hailed from south and central Wales and Monmouthshire. At Bangor in 1958, on the other hand, it was noticeable how very poorly the breeders from the southern and western counties of Wales were represented. Indeed, in some of the classes at Bangor, entries from over the border outnumbered those from Wales! Among the thirty-two Herefords judged, none of the prizewinning herds of the southern counties was represented and more than half the entries came from beyond Offa's Dyke. There was an absence, too, of entries from the Shorthorn breeders of the south and west of Wales, and despite the fact that they were at this time the most popular breed in Wales the Shorthorn competitions were almost dominated by English breeders. British Friesians made a good show with sixty-three entries, but they were mainly from north Wales.

Welsh Black cattle – especially those in the south-western counties – had in the years down to the 1930s lost ground to Shorthorns, Herefords and Friesians, though they happily recovered some of the lost ground in the 1940s. While Dairy Shorthorns were still the most popular breed in Wales to emerge from the war years, other breeds were by then growing in importance, notably the Ayrshires and Jerseys. These alterations in the breed structure were, as we would expect, reflected in the growing entries of the newcomers in the various cattle sections at the Royal Welsh Shows. Welsh farmers at the Abergele show in 1950 were thus rueful about the rather poor lot of Welsh Blacks on show compared with the magnificent entry of more than a hundred Ayrshires of outstanding quality and spoke feelingly of 'the Scottish invasion of Wales'. It was plain to see that their deep milking qualities were making a strong appeal to the Welsh dairy farmer. Despite these several 'invasions' by outside breeds from the start of the century, the native Welsh Black cattle,

Table 3. Cattle entries at Royal Welsh Shows, 1947–1962

BREED	1947	1948	1949	1950	1951	1952	1953	1954	1955	1956	1957	1958	1959	1960	1961	1962
Welsh Black	71	–	77	65	87	–	48	89	73	86	130	88	49	67	76	75
Shorthorn	99	–	88	60	126	–	73	63	71	72	40	27	33	44	52	63
Hereford	31	–	41	45	67	–	66	50	52	49	37	32	47	71	53	35
British Friesian	15	–	28	41	71	–	49	72	79	88	50	63	43	86	79	53
Ayrshire	52	–	52	92	90	–	53	79	62	91	55	45	39	69	51	47
Other cattle	–	–	–	31	63	–	54	76	92	70	57	27	64	–	–	–
Jerseys	–	–	–	–	–	–	–	–	–	–	–	–	–	62	40	36
Guernseys		–	–	–	–	–	–	–	–	–	–	–	–	23	23	41
Aberdeen Angus	–	–	–	–	–	–	–	–	–	–	–	–	–	11	12	6

though increasingly confined to the north-western areas, had their followers in all other Welsh counties, and improvements in the breed came from the early century largely through the efforts of the Welsh Black Cattle Society and the Livestock Improvement Scheme and the inspired guidance of Moses Griffith with his choice herd at Egryn, Merioneth. Reference to Table 3 will reveal that at the shows from 1957 to 1962 the native Welsh Black cattle had indeed recovered their position as the largest cattle section on display.

As had always fittingly been the case at this national show, Welsh Black cattle occupied the premier position in the main ring Grand Parade of winning cattle in the post-war years, and the 'Battle of the Blacks' was eagerly anticipated by agricultural visitors to the show. Continuing his run of success at the pre-war shows, a highly successful Royal Welsh Show exhibitor in the Welsh Black classes at both Carmarthen in 1947 and Swansea in 1949 was J. M. Jenkins of Tal-y-bont, the 'grand old man' of the breed, who brought to Swansea the two Welsh Blacks which had taken the laurels at the recent Royal Show at Shrewsbury. His fine bull Neuadd Idwal not only won four firsts but also the Colonel Harry Platt Memorial Challenge Cup for the best exhibit in the Welsh Black cattle classes, while his cow Caran Tilly, in addition to winning several firsts, gained a championship award given by the Welsh Black Cattle Society for the best female in the show. Carrying all before him in the early 1950s as winner of the championship at four Royal Welsh Shows was the famous Welsh Black bull Egryn Garnedd, bred by Moses Griffith and owned by Richard Rees of Ynys Farm, Machynlleth. His prowess compared favourably with that of the champion bull of the Aberystwyth shows between 1906 and 1908, Duke of Connaught. Of course, age catches up with the best, and the first jolt came at Haverfordwest in 1955 when this 'uncrowned king' of the Welsh breed was sensationally beaten by his youngest daughter Ynys Glenca 3rd, bred by the same Richard Rees. Even worse humiliation would

35. J. M. Jenkins, Taliesin, at the Carmarthen show, 1947, receiving the Prince of Wales Challenge Cup from Princess Elizabeth for the best group of four Welsh Blacks.

be visited upon the 9½-year-old bull at Rhyl the following year when he was defeated by Ysbyty Ifor, a three-year-old exhibited by Gwilym Edwards of Bala. Indeed, for the first time at a Royal Welsh Show he was forced to take his place at the bottom of the line in the judges' ring, a self-confessed 'blow' to Rees who believed that the old champion was good enough for second or third place; 'the bull was trotting around like a pony this morning and showed little signs of age', he complained. Ysbyty Ifor was to establish his supremacy by winning the championship at Aberystwyth in 1957, but upon his sale to the

Milk Marketing Board his throne became vacant at the Bangor show in 1958. Emerging victorious from among the eighty-nine contestants was the three-year-old bull Rhyllech Cymro, owned by William Owen of Pontfadog, Glynceiriog. Over the years the Royal Welsh Show has been packed with little human dramas of one kind or another and in this 'Battle of the Blacks'

36. The famous Welsh Black bull, Egryn Garnedd.

at Bangor the winning bull would not have appeared on the showground at all but for the determination of William Owen's son David to bring him there on his own from the family's remote farmstead in the face of his father's indisposition and consequent decision to withdraw the entry. Nor would his father have learned the good news until he got the newspaper at 5 o'clock the following day. By contrast, the glad tidings that his bull Esgob Emrys 2nd had won the Breed Supreme Champion Award were immediately telephoned from the Gelli-aur show in 1961 to the unavoidably absent Richard ap Simon Jones of Ysguboriau, Towyn, by his herdsman. With his owner the next time standing with other Welsh Black enthusiasts for hours in the drizzle around the ringside, he repeated his achievement at Wrexham in 1962.

Classes for other breeds attracted prominent exhibitors in post-war years. Doubtless a source of considerable satisfaction to the Society's organizers was the increase in numbers of entries in the British Friesian classes after the scant support accorded by breeders in the inter-war years. John Bennion of Stackpole Court, Pembrokeshire – particularly with his winning bull Stackpole Engelsham 2nd at Machynlleth in 1954 – E. C. E. Griffith of Plas-newydd, Trefnant, Denbighshire, E. M. Corfield and Son, Great Moat, Montgomeryshire, and C. E. B. Draper and Son, of Acton Burnell near Shrewsbury, were notable breeders who gained success in this section, but the most dominant among them was Montgomeryshire breeder, R. W. Griffiths of Forden. The most impressive of all his exhibits was his famous bull

Holmside Sure, which carried off prizes at the Royal Welsh and elsewhere, including the Royal, in the late 1950s. The Dairy Shorthorn section witnessed a remarkable success for the duke of Westminster's Eaton herd at the 1950 and 1956 shows, with ringside onlookers at the latter event witnessing as breed champion Eaton Wild Eyes 111th, a dairy cow of very great merit with exceptional udder formation. T. Llywelyn Jones of Ystrad Farm, Carmarthen, did well with his beef Shorthorns, winning, for instance, at Haverfordwest in 1955 the principal trophy with his bull Barton Silver Ace. Exhibitors in the Hereford classes included an outstanding breeder – W. E. Thorne of Studdolph Hall, Milford Haven – whose beautiful ten-month-old calf Studdolph Mabel shown at Aberystwyth in 1957 was probably the first Hereford heifer ever to win the championship at such an early age. Other prominent winners of Hereford classes were W. Milner of Much Wenlock, Salop, whose young twelve-month-old bull of his own breeding, Wenlock Gringo, for example, took the supreme championship of the breed at Rhyl in 1956, and T. L. Parker of Bishop's Frome, Worcestershire, who in 1961 became entitled to keep the Tarrington Challenge Cup after winning it three times. Ayrshire classes, too, attracted some prominent breeders, notably W. H. Slater of Wellington, Shropshire, W. Craven Llewelyn, Cefn Cethin, Llandeilo, John Bourne, Moreton-in-Marsh, and G. H. Dodd and Son, Ellerton Grange, Newport, Shropshire. Prominent among exhibitors in the Jersey classes in the mid-1950s was R. F. Wynne of Rhuddlan, whose prize-winning six-year-old bull Jingo's Spoilt Boy at Rhyl in 1956 was a magnificent specimen. A notable exhibitor of the breed towards the close of the migratory show era was Harold Embrey of The Brooks Farm, Raglan, whose bull Abinger Harmonie won the Show Society Championship Rosette for the best male exhibit in three successive years down to and including 1962.

Just as Welsh Black cattle caught the eye of showgoers, so did the Welsh breeds of sheep, some of them quite recent additions stretching back no further than the Second World War. Welsh Mountain, numerically the most important, South Wales Mountain, Llanwenog, Beulah Speckled Face and Black Welsh Mountain were all promoted by their own societies, and the Royal Welsh Agricultural Society also played a leading role in improving these hill flocks. Other sheep popularly found in Wales were Clun Forest and Radnor, whose classes were also well-filled at successive shows. At the Aberystwyth show in 1957 the famous six-year-old Clun Forest ram, Court Llacca F.50, was shown by the celebrated exhibitor of the breed over these years T. R. Eckley of Felin-fach, Breconshire, and won for him the male championship of the breed. Such indeed was the ram's prowess that he won no fewer than thirty-two championships, including two Royal, two Bath and West and two Royal Welsh championships! A notable Royal Welsh winner in the Welsh Mountain sheep classes in the later run of migratory shows down to 1962 was John Ellis Jones of Blaen-y-cwm, Llangynog, near Oswestry, who won all the special prizes and two firsts and one second of the class prizes in the Hill Flock Section of the breed at Welshpool in 1960. This was his most successful Royal Welsh to date, though he had won the championship ram

37. Clun Forest ram, Court Llacca F.50.

twice before. Impressively, he again won the award for the champion hill flock Welsh Mountain ram at Gelli-aur the following year with a different animal from the one shown at Welshpool. His flock of almost 2,000 hardy Welsh Mountain sheep, kept by him and his two sons on his 3,000-acre farm in the Berwyn mountains, was originally founded by his grandfather, who had moved to Blaen-y-cwm in 1883 when the Vyrnwy valley was flooded.

Although horses and cattle were the traditional eye-catchers, one commentator on the 1954 show at Machynlleth claimed that perhaps the most 'spectacular feature' on the first day were the Welsh pigs, of which 140 entries paraded the show ring. Having already carried all before it at the recent Royal Show in Windsor, the fine boar Musslewick Supreme 2nd, belonging to Malvern Farms Ltd who had purchased it for the record figure for any pig in Great Britain of £525, took the supreme championship of the breed at Machynlleth. Despite going on to win the supreme championship at the Royal Show the following year, his entry in the Welsh Pig class at Haverfordwest shortly afterwards saw him beaten into third place by Teilo Solomon 4th, a boar bred by Laurie Evans of Penally (Pembs.) and shown by W. Murray of Eastington Farm, near Pembroke. His first time of showing, Murray had a very successful day in bringing six pigs to the show and winning eight prizes. At the 1956 show in Rhyl, Welsh Pig entries were commended for their great improvement in quality, especially the gilts. A young gilt, Letton Lunette 2nd, won the supreme championship for S. S. Eglington and Son of Thetford,

Norfolk. The best boar was the three-year-old Teilo Solomon bred by the aforementioned Laurie Evans, from the famous Temple Druid Acorn 3rd and shown by J. M. Whelan of Merlin's Bridge, Haverfordwest. It was a nice compliment to the popularity of the Welsh breed that the majority of the prizes in the Welsh Pig Section at Aberystwyth in 1957 went to English

38. Welsh sow, 11225 Letton Lunette 2nd.

breeders. The Eglington Cup for the best exhibit in the Welsh Pig classes went for the second year running to the donor, S. S. Eglington, whose campaigning efforts since the war had led to the breed, after losing much ground in the 1940s, becoming such a prominent feature of Welsh animal husbandry by the late 1950s. Following the famous Peterborough sale in 1953 of Landrace stock imported from Sweden, the Haverfordwest show in 1955 brought to the public eye the potential of the Landrace pig as the perfect bacon pig. The three 'local' Landrace classes were won by three Pembrokeshire brothers, David C., R. G. N. and George G. Llewellin, farming independently of one another. Landrace open classes were introduced for the first time at the Rhyl show in 1956.

For all the excitement of the livestock rings and the Grand Parade and other displays in the main ring, the latter graced by a high standard of sartorial excellence on the part of the stewards with their smart bowler hats, immaculate suits, shooting sticks and the club or regimental ties, the avenues of machinery were always thronged with people interested in new or improved machinery and implements. The war years had given the initial impetus to mechanization because of the need to plough up as much land as possible; the tractor, in particular, proved itself indispensable to cope with the increased tillage acreage. Its advantages thus soon became apparent to the Welsh farmer. Not surprisingly the 100,000 square feet allocated to agricultural machinery at the Swansea show in 1949 exceeded all previous space records. Of course, this was a growing department of the show, and one commentator later observed that the 1960 show at Welshpool boasted one of the most varied and comprehensive displays of farm machinery ever seen at the Royal Welsh. Besides tractors, a wide range of new machinery was now coming to the farmers' aid. The pick-up baler became very popular throughout Wales in the early and mid-1950s and among the orders placed in machinery avenue of the Cardiff show in 1953 balers were in biggest demand. Their latest models were exhibited at Rhyl in 1956. Improvements and additions to the numerous combine harvesters, too, were seen at that same show, such as the straw press attachment in the Massey Harris combine.

The Royal Welsh Agricultural Society, through its Machinery Committee founded in 1951, also organized competitions both at the show itself and some conducted apart from the show. We have already heard about the innovative Silver Medal Competitions for new machines or implements or adaptations which were likely to make the most useful contribution to meet Welsh upland farming conditions. Working exhibitions in connection with these machines, suited to Welsh upland terrain, were also staged at the show from 1951 onwards, but in 1955 the Machinery Committee deplored the lack of enthusiasm on the part of the manufacturers and expressed disappointment at the number of entries received over the last few years. It was felt that a certain competitive element attached to the scheme had made it unpopular and it was therefore decided to switch to staging machinery demonstrations on a particular theme. All the manufacturers within the operation were invited to participate in practical demonstrations on a site adjacent to the showground which could be staged during the show. Thus at Rhyl in 1956 ditch maintenance demonstrations were organized. However, it was recognized in 1961 that the efforts of the Machinery Committee to obtain the support of manufacturers in this direction had not borne fruit. It was hoped in 1962 that the National Institute of Agricultural Engineering would be able to stage a practical exhibit at the 1963 show and at subsequent ones.

In a very real sense the Royal Welsh comprised many shows within the main show. This was certainly true as far as the educational exhibits were concerned from its inception in 1904 onwards, and an even higher profile was achieved in the years from 1947 to 1962. Perhaps the outstanding feature of the Society's annual show in these last migratory years was the huge emphasis placed on the educational exhibits arranged on the showground by the many and various organizations and educational departments. We have alluded earlier to the 'new departure' in 1960 at Welshpool in the form of extended space devoted to educational exhibits, some eight acres in all, and at Gelli-aur the following year yet more attention was paid to practical demonstrations and educational exhibits. All this would have met with the approval of Sir

C. Bryner Jones (d.1954), who, in his lecture to the Society on 12 July of that year, emphasized: 'But holding the Show is not the only purpose of the Society and competition as such is not the only purpose of the Show. There is an educational aspect to the Show and one of the most important aims of the Society is to give the farmer – the small farmer as well as the big farmer – the opportunity to see what scientific research can do to throw light on treatment of the land, breeding and feeding animals, improving pasture and all other aspects of farm work.' Prominent over these years from 1947 onwards were the educational exhibits of the Ministry of Agriculture (NAAS), which addressed such themes as milk and milk production (1949), grass and sheep (1950), increased meat production, especially bacon (1953), rearing of young animals (1956) and improving the Welsh Mountain ewe (1958). Always a popular calling place on Educational Avenue, the Ministry's 1950 stand thus demonstrated to sheep breeders by way of examples of Welsh ewes mated to Down rams how, on improved pastures, these sheep provided enough milk to fatten lambs of prime quality. Other educational and demonstration stands included those of the various Marketing Boards – Milk (whose 1955 Haverfordwest milk stand was allegedly the busiest on the showground), Egg, Wool and Potatoes. Particularly attractive was the stand of the Wool Secretariat, especially during the times of the mannequin parades from 1956 onwards. In the growing and initially uneasy partnership between farming and afforestation, the Forestry's stand was always instructive in its coverage of the

39. Ministry of Agriculture exhibit at the Cardiff show in 1953.

40. Women's Institute cake stall at Machynlleth, 1954.

many aspects of forestry work, including 'live' demonstrations. Valuable, too, were the extensive programmes of educational competitions organized by the Young Farmers' Club movement at successive shows. The Society gave an annual grant aid of £150, together with free space, free printing and equipment, to help it stage its comprehensive exhibit and hold its numerous educational competitions. Its exhibit at Rhyl in 1956, for instance, took the theme 'From the Land to your Table'. Free site and shedding were also placed at the disposal of the Rural Industries Bureau which organized an extensive programme of educational competitions, and also a textile section. Likewise, free space was provided for the Women's Institute exhibit, closely related to the Produce Section of the show, which in 1958 at Bangor, for instance, took as its theme 'From Print to Practice'. Besides competitions held within the Produce Section at the show, demonstrations were also staged. Following discussions of the Society's Produce Demonstrations sub-committee in 1960, it was decided that future demonstrations at the show should concentrate on one theme, which covered butter and cheese making in 1961 and a home-freezing demonstration in 1962. Other educational activities connected with the show were the informative show-ring commentaries, competitions to encourage better skills, as the sheep-shearing ones from the close of the 1950s, and the staging of educational competitions, for example – from the start of the 1960s – those of lamb and bacon carcasses. In 1954 Sir Bryner was indeed to point to the educational side of the Society's work as having contributed significantly to improvements that had taken place in Welsh farming since the start of the century: 'Since the foundation of the Society in the year 1904, many changes have taken place in Welsh agriculture and doubtless a great deal of improvement in particular directions. It could be said that the art of agriculture is of a higher standard today – although not everywhere – than it was fifty years ago, and the Royal Welsh can claim some of the credit for this because of the prominence it has given down the years to the *educational aspect* of its work.'

Although the Young Farmers' Clubs joined the Royal Welsh scene before the Second World War, their involvement was necessarily limited, given that in 1939 there were only forty clubs in Wales. This number grew to 124 by 1943 and mushroomed thereafter. It was not surprising, therefore, that it was after the war that the movement came to play a big role in the Society's affairs. Aware of their importance for the future, the Society now granted them their own self-contained 'enclosure' at the show, a privilege subscription was extended to their members and £200 were allocated to them each year, a sum which was reduced to £150 in 1957. In 1946, too, the future secretary of the Royal Welsh Agricultural Society, Arthur George, successfully submitted a number of recommendations to the Society in connection with competitions at the 1947 show and towards gaining for the Young Farmers' Clubs co-opted membership of the Council and the Stock Prizes and Judges Selection Committee. This meant that, subject to the Young Farmers providing the prize money, appointing their own judges and stewards and making arrangements for the organization and management of their competitions, the latter would be accepted as an integral part of the show programme in the years to come. The Young Farmers' Club Section had veritably become 'a show within a show'. The competitions were concerned with the practical aspects of farming and of home life. In relation to the latter category, the international competition in demonstration of craftwork was introduced at the Rhyl show in 1956. Stewards and judges of the various competitions were men and women who had themselves come up through the Young Farmers' Clubs and would later become stewards and junior judges in the livestock section of the show. As shown above, a special exhibit at the show was a contribution by the Young Farmers themselves. Moreover, Young Farmers did not make just for their own enclosure and stay within it! Many clubs organized a special educational trip to the show to inspect all the sections, the research and experimental plots as well as the exhibits mounted in the Educational Avenue. The sheer growth in the Young Farmers' presence can be seen by comparing the Carmarthen show in 1947 and that of Gelli-aur in 1961: whereas at Carmarthen the Young Farmers' corner of the show was a modest affair – merely a small canvas stand with limited space and rather a modest display of products – at Gelli-aur their enclosure was the largest to date, and the various national (all-Wales) competitions were keenly contested by the different Welsh counties. The contests at Gelli-aur included poultry judging and trussing, dairy, beef and Welsh Black cattle judging, sheep shearing, and special competitions for girls in floral decoration, decorating cakes and trimming hats.

The Women's Institute movement in this post-war era strengthened the links it had already begun to forge with the Royal Welsh Agricultural Society before 1939. In late 1954 the National Federation was thanked most warmly for its 'wonderful support' at the Machynlleth show earlier in the year, while later, in 1960, its excellence, especially in educational matters, was commended. They were such staunch allies that in October 1961 the Welsh organizer of the National Federation of Women's Institutes, Beti Jones of Newcastle Emlyn, was co-opted onto the Society's Council. When Margaret Brodrick, a previously elected member of Council for Denbighshire, died in 1962, Mrs Jones joined Jane Davies of Felin-fach, Lampeter, a co-opted member representing the National Federation of Young Farmers' Clubs (since 1950), and Mrs N. Pennell of Hartpury, Gloucestershire, an elected member for the Outside Counties, as the only female members of Council. Not only did the Women's Institute organize its own stand and exhibit at successive shows, but local branches were invited to supply the required number of stewards in the Produce Section of the show and they also played a part in door-to-door collecting for the show's local fund.

The Horticultural Section was also very much 'a show within a show'. Some difficulty was experienced under the migratory system in attracting large-scale support from national horticultural traders when the show was held in remote areas. Consequently the Floral Section down to the mid-1950s was unsatisfactory. Growing awareness of this led the Society in late 1955 to decide that, in future, traders of national standing would have to be invited to exhibit very much earlier than hitherto and that the prize money would be significantly increased. Prompt implementation of these resolutions yielded results, for the number and quality of the exhibits at the Rhyl show in 1956 were a great improvement on previous years. But the position of the flower show became uncertain in the financial crisis facing the Society later in that year and, because of doubts felt, in the light of past experience, that exhibitors of national repute would travel to the show at Aberystwyth, the flower show was abandoned for 1957. Whilst it was not possible to stage a dog show the following year at Bangor, happily the Horticultural Committee was allowed to put on a flower show there, although it was to be on a self-supporting basis or as near to this as possible. After these difficulties, Society officials would have been pleased by the outstanding success of the Floral Section at both Gelli-aur in 1961 and at Wrexham the following year.

Certain aspects of the show were crowd-pullers and appealed to country dwellers and townspeople alike. Sheepdog trials for the Royal Welsh Stakes had acted as a big draw in the years before 1939, and Captain Howson, in describing the Swansea show of 1949 – where forty-two competitors entered for the Royal Welsh Stakes – commented that they appeared to increase in popularity year upon year. Their crowd appeal notwithstanding, the honorary director, Lieutenant Colonel FitzHugh, suggested in 1955 that sheepdog trials might sensibly be staged apart from the show in an area distant from the show centre, on the grounds that this would allow for a reduction of fifteen to twenty acres in area required for the annual show. More sites in Wales would thereby become available for the show and the arrangement would also serve to keep the interest of the Society members in those parts of Wales where the trials were held, members who, because of distance, might not be able to attend the annual show that year. Certainly in the absence of suitable space on the show site at Rhyl in 1956, the sheepdog trials were not held as part of the show that year. Likewise, because a minimum of 400 yards was required to stage them, no sheepdog trials were held at Bangor in 1958. A pattern of cancellation was setting in, and when a request was received in September 1961 from the South Wales Sheep Dog Association for the revival of the Royal Welsh Sheep Dog Trials, Lieutenant Colonel FitzHugh pointed out that this was an impossibility at Wrexham but that the matter would certainly be considered for the Builth Wells show in 1963.

Displays in the main ring were as attractive in these years as they had been in earlier ones. Despite the Society's policy in this period of keeping the cost of main ring displays down, thus ruling out spending sums like £1,000 to £1,500 on special displays, the events were varied, colourful and often – as was the motorcycle display at Margam in 1959 given by the display team of the 31st Training Regiment, Royal Artillery – spectacular. Prominence was always given to the show jumping, which consistently reached a high standard. Margam show in 1959 saw several riders of international fame come down from the White City, London. Their prowess only served to emphasize the achievement of local riders like David Broome of Chepstow and Pat Price in winning their competitions. In the years following the end of the war military displays were popular, such as that given by the City of Liverpool

41. Liverpool mounted police at the Royal Welsh Show in Caernarfon, 1952.

Mounted Police at the opening of the 1950s, but the customary mounted police display was superseded at Machynlleth in 1954 by special parades of Welsh Mountain ponies, Welsh cobs and Welsh ponies, arranged mainly for buyers from abroad who were visiting the show. As well as the twenty-four horses and riders of Tunstall Hall College, Market Drayton, and the motorcycle regimental display team of Kinmel Park Camp who entertained onlookers in the main ring at Rhyl in 1956, sheepdog demonstrations were also a feature, doubtless on account of the cancellation of the sheepdog trials at the show. Similar demonstrations occurred at later shows, like the one given by Meirion Jones of Llandrillo, winner of the 1959 International Sheepdog Trials, at Welshpool in 1960. Although foxhounds played a diminishing part in the shows after the war, parades of them continued to add colour to the main ring as, too, did the familiar ceremonial displays of the military bands.

At one of the shows held between 1947 and 1962 the main ring display was a demonstration of aerial crop spraying by helicopter. Graham Rees, who became chief steward, main ring, in the last years of the twentieth century, recalls how the occasion descended into farce. In his toings and froings at a height of about thirty or forty feet, the helicopter pilot caused commotion on the ground by scattering all over the ring the papers of commentator, Percy Thomas, who was seated in the announcer's box – comprising a small rostrum five feet square in the centre of the main ring – and blowing about people's coats, trousers and bowlers. In its final triumphant flypast the helicopter sprayed water all over the main ring stewards, at which point Major Jack Rees, the chief steward in the main ring, expostulated: 'Get this . . . out of here!' Although coming through loud and clear on the PA system, the colourful language of the instruction is unprintable, Rees assures us.

A new sentimental focus of the show in these post-war years was the award of long-service medals to farm and other rural workers. At Swansea in 1949, proudly watched by their employers Major and Mrs Gibson-Watt of Doldowlod, Llandrindod, and forty fellow employees on the estate, three workers were presented by Field Marshal Montgomery with the medal of the Royal Agricultural Society of England for long service on the land. Early 1956, however, saw the Royal Welsh Agricultural Society inaugurate its own Long Service Medal Scheme. It was felt that the conditions of the RASE award were unsuitable for Welsh conditions and the National Farmers' Union fully supported the initiative. Twenty-nine winners received their medals at the 1956 show in Rhyl. The Society's official report on the Aberystwyth show the following year claimed that probably what would remain in the minds of many was the long line of old agricultural workers who had qualified for the Society's long-service medal, especially because 'a worthy farm maid', Miss Alice Jane Jones, was at the head of the column.

As in the years before 1939, so too in the post-war era a planned visit by a member of the royal family was eagerly anticipated and drew large crowds. On no occasion was this more evident than during the visit in her capacity as the Society's president of Princess Elizabeth to the Carmarthen show on its opening day in early August 1947, when she was ecstatically greeted at every stage of her progress. (Later, in 1952, as Queen Elizabeth II she would intimate her intention of becoming patron of the Society, following in that office her father, King George VI, and her grandfather, King George V.) Likewise, Princess Alexandra, who visited the 1957 Aberystwyth show on its second day, won the hearts of what was a record second-day 40,000 crowd by scorning protocol and insisting on mingling with the farming people during her five-hour walk about. A jubilant show official pronounced on this marvellous second day: 'If it does turn out to be a record-breaking Show, we'll only have one person to thank – Princess Alexandra.' Unfortunately, that Thursday night a cloud-burst in the upper reaches of the River Rheidol coincided with a high tide at its estuary. When the two strong currents met, the river overflowed its banks and flooded the showground, so that it became virtually an inland lake during the Friday morning of the show. This meant that the princess's official programme had to be curtailed. If a record attendance over the three days was not thereby achieved, a good number of visitors braved

42. Princess Elizabeth at the Carmarthen show in 1947.

the conditions on the third day. Such was the rapid sale of Wellington boots early in the day that all the available stocks in the neighbourhood were cleared. Pugh James, who kept a shoe shop southwards in Aberporth, having heard about the flood, set off early on the Friday morning with all the Wellington boots he had in stock. Upon his arrival, he found that conditions were so bad on the showground that people began bidding for the 'Wellies'. All were sold in a matter of minutes and, having also sold the ones he was wearing, he had to drive home barefooted!

As in the years before 1939, one of the social events of the annual show was provided by the overseas visitors who attended as the Society's guests and who were in these years, by arrangement, under the stewardship of the Wales Tourist Board. Located in their own Overseas Pavilion, they came from a wide range of countries including Canada, Australia, New Zealand, Jamaica, Rhodesia, Kenya, Pakistan, the United States, Austria, France, Norway, Holland, Sweden, Denmark and Yugoslavia. Indeed, those visiting the Welshpool show in 1960 came from thirty different countries. While many were farmers visiting Britain specifically to study its agricultural practices, as, for instance, were the six young men from Yugoslavia who came to the Haverfordwest show in 1955, some, especially from the United States and Canada, were farmers who had left Wales to till the soil overseas. One such emigré Welshman was Jesse Roberts of Hannah, Alberta, a visitor at the Llandrindod Wells show held at Llanelwedd in 1951, who, with his brother Tudor, had left the neighbourhood of Diserth in Flintshire in 1907 to carve a

43. The Aberystwyth showground under water in 1957.

successful agricultural career out of tough marginal land in Canada. Up to a hundred overseas guests were on occasions to be found at the Royal Welsh Show during these years.

Despite the adverse publicity given the Bangor show of 1958 by the press – perhaps because of the terrible conditions underfoot – which allegedly reduced attendance on the last two days and understandably angered the Society's officials, newspapers continued to play a valuable role in promoting the annual show. Not just the Welsh ones, but others like the *Daily Telegraph*, the *Daily Mail* and the *Birmingham Post* covered various shows; so grateful was the Society to Percy Izzard of the *Daily Mail* for his long association with the Royal Welsh that on the occasion of his last show at Machynlleth in 1954 he was granted life membership of the Society. On the death in 1958 of W. J. Jones, who had reported on the Royal Welsh Show for a number of years for the *Birmingham Post*, his colleagues at the *Post* presented a trophy as a perpetual award in his memory to the Royal Welsh Agricultural Society to be allocated to the Hereford Section. Singled out by the Society's officials for special thanks for its extensive coverage of the Cardiff show in 1953 was the *Western Mail*, and deservedly so in the light of its resumption the previous year of its pre-war practice of publishing valuable agricultural supplements in the run-up to the annual show. Chief agricultural journalist for the paper in the 1950s was Hugh Busher, 'Landsman', who was to be joined in 1960 by Roland Brooks.

Of great and growing importance, of course, in the post-war years was television coverage of the show. Television coverage began with the

Haverfordwest show in 1955, where it took place on the first evening. Notwithstanding the generous time allotment, the programme was criticized by 'Y Tiwniwr' in the *Western Mail* for including a rather dull Ministry of Agriculture film on farming development in west Wales at the expense of more show 'actuality', which could have been achieved by cameras ranging still more widely over the main ring and bringing in as many lively personalities, people and animals alike, as possible. Television coverage thereafter was on a fast learning curve, and a major new feature would come in 1961 with a colour broadcast of the 1961 Gelli-aur show.

A towering personality in organizing the show and ensuring its success year after year in the post-war era was the main ring senior steward, Alan Turnbull of Gower. The secretary Arthur George later paid handsome tribute to him in his private papers: 'The greatest "find" for the Society and to me personally, as I gained a friend for some forty-three years, was the late Alan Turnbull.' Not that George was blind to his friend's fiery temper; he recalled the occasion at the Haverfordwest show in 1955 when Turnbull stalked out of the main ring as a result of a 'little disagreement' over the order of the main livestock parade. Fortunately, George saw his long strides taking him towards the car park in his determination to leave for home. Persuaded to have a cup of tea in the Georges' caravan, he soon cooled down and within a short time was back in the main ring, the differences already forgotten! His chairing of the all-important Main Ring Committee, often meeting at the Castle Hotel, Neath, was always a pleasurable occasion, recalled George.

In autumn 1960, with the decision to move to a permanent site already taken, the question was asked: 'Should the annual show continue to occupy the paramount position in the Society it now does?' After full discussion it was decided that any planning for a permanent site should proceed upon the basis that the show should remain very much as at present and approximately at the same size. Moreover, it was resolved that amenities, both for members and the public, should be improved; that live demonstrations, in which the Society had already made some progress, should, wherever possible, be extended to other sections of the show; and that stock sales should be included, in the same way as was done in the Dublin show. We have seen that the Society also gave serious consideration from 1960 onwards to the future pattern of livestock classification at the show because it had realized that the current pattern did not meet the requirements of the industry as represented by progressive breeders and commercial farmers. It was accordingly resolved that some modification was necessary, with one immediate innovation taking the form of carcass competitions from the 1961 show onwards.

CHAPTER NINE

Pastures New

The 1950s was a decade of prolonged introspection for the Society into its role within Welsh agriculture and the rural community. This was prompted by a memorandum prepared by Moses Griffith in 1951. Basically he wished to see the Society play a more effective part in influencing progressive agricultural policy in Wales and thereby raise its status. While recognizing that the Society's chief function to date of holding an annual show was a 'very worthy object', he believed that as a national body its objectives and functions should be much wider. Accordingly, he suggested that the Council should nominate a working party to study and report on: the provision made for agricultural education and research in Wales; land utilization in Wales; and the development of native breeds of livestock. In response, in May 1951 the Council set up a working party to examine the Society's objectives, to report how far these objectives had been attained, and to recommend further courses of action. Those nominated to the working party were Lieutenant Colonel O. W. Williams-Wynn (chair), D. S. Davies, J. E. Gibby, R. L. Jones, Professor J. E. Nichols (Aberystwyth), Professor E. J. Roberts (Bangor), and Moses Griffith. After no fewer than ten meetings, the working party submitted its recommendations to Council in March 1953, along with a minority report by Moses Griffith, who, among other recommendations, pressed the need for a bilingual *Welsh Journal of Agriculture* and for more use of Welsh in the running of the Society's activities if it was to win greater support from the Welsh farming community.

Comprehensive in their range, the recommendations of the majority report, which were accepted by Council in March 1953, fell under a number of headings. At the top of the list was the provision of agricultural education in Wales: the working party called for the establishment of a 'two winter session' diploma course in agriculture at farm institutes in Wales in addition to the one full-year course currently offered. Other recommendations included holding an annual conference; setting up a hill-farm research station in Wales; the revival of the *Welsh Journal of Agriculture*; organization of demonstrations in the grading of animals and their carcasses, in conjunction with the annual show; establishment of a National Institute of Agricultural Engineering sub-station in Wales; implementation of an essay competition; extension of the existing cooperation between the Society and the Ministry's National Agricultural Advisory Service; organization of tours of overseas countries; instituting a Gold Medal Award in recognition of any outstanding service in the field of practice or science given to Welsh agriculture; organization of architectural competitions in view of the great dearth of expert advice available to the farmer and the difficulty in obtaining adequate improvements to existing buildings or the erection of suitably planned new buildings of good architectural design; institution of a Long Service Medal Scheme; circulation of a full annual report to members before county meetings were held in the spring of each year to counter the lack of periodical reports to members of the Society's activities; support for the University Court's decision to establish a veterinary school in Wales; and, finally, the grasping of every opportunity to extend cooperation between the Society and other bodies with similar interests. Additional recommendations were made by the working party following a further memorandum submitted by Moses Griffith in 1957 emphasizing that the relationship between the Society and the ordinary Welsh farmer had been 'far too lax and ineffective' and that the Society should endeavour to provide facilities for Welsh agriculture similar to those that had been won in England under the aegis of the Royal Agricultural Society and in Scotland under the Royal Highland Agricultural Society.

Reviews on progress towards implementing the working party's recommendations were made for the Council's benefit by the secretary, Arthur George, in November 1955 and again in March 1958 and by the newly formed Education Committee still later in March 1962. In so far as diploma courses in agriculture and dairying in Wales were concerned, the Ministry of Agriculture, in response to the Society's call in 1953 for a review of the position, appointed in December 1955 a committee chaired by Professor D. Seaborne Davies, which heard oral evidence from Society representatives. When the committee's conclusions were published in autumn 1956, they were welcomed by the Society as they largely coincided with the views submitted by their representatives. The committee's major recommendation was the establishment of a new agricultural college at Aberystwyth, whose purpose would be to provide facilities to train students to diploma levels in agriculture and dairying, as well as to be the centre for a forestry school. Since a new agricultural college was believed to be essential to the future prosperity of the industry, this recommendation was strongly supported by the Society among other public bodies in Wales – notably the Welsh Joint Education Committee. A deputation of the Society's Education Committee subsequently appeared before Welsh MPs at Westminster and later before the Council of Wales, the body which in autumn 1962 would recommend the establishment of an Agricultural College for Wales. The Education

Committee, in January 1963, also took pleasure in the news that, in response to the approach of the Royal Welsh Agricultural Society, Aberystwyth had decided to extend its dairying diploma course until autumn 1963.

Following the appointment in late 1959 of an ad hoc Education Committee to review the Society's educational activities, a deputation headed by T. H. Jones appeared before the Agricultural Research Council in London in January 1962 to press the Society's support for the establishment of an Animal Health Research Station in Wales. Sadly, the meeting was marred by the collapse and death of R. W. Griffiths, a member of Council since 1954 and a vice-president of the Society since 1960. Although Lord Hailsham's ruling, as minister of science, in spring 1962 that agricultural science would be better fostered within the University of Wales rather than by the creation of a new station came as a disappointment to the Society, the campaign had concentrated minds on the need for action in the face of the heavy annual losses of livestock in Wales.

The working party's report also led the Society to call upon the National Institute of Agricultural Engineering to establish a sub-station in Wales to cater specifically for the machinery requirements of Welsh hill farmers. In the course of pursuing this objective, cooperation between the Society and the Institute eventually led in 1958 to an invitation to the Society to submit the names of three suitable persons to represent it on the committee of the Scottish sub-station. Colonel J. J. Davis of Llanina mansion, New Quay, was duly appointed by the governing body of the Institute to represent the Royal Welsh Agricultural Society on its Scottish committee, a body he joined in April 1958. Even if the Society's aim for a sub-station to be established in Wales was not achieved, its representation on the Scottish committee secured engineering data in the interests of Wales.

We have already seen that the working party's recommendation of a Long Service Medal Scheme was implemented in 1956. So too, in the same year, was the recommended architectural competition based on the replanning of a farmstead, a competition adopted by the Machinery Committee and which owed much of its success to the enthusiastic guidance of Leonard Williams of the Agricultural Land Service.

An annual conference discussing an important topic relating to Welsh agriculture was first instituted by the Society in spring 1954 at Aberystwyth. It took as its theme 'The Problem of Home Produced and Imported Meat'. Similarly, as a further attempt to improve the services that the Society gave its members and in this instance by enabling them to see what other countries were doing to face problems that could also affect the Welsh farmer, in 1953 a tour was organized to Holland and in the following year to Denmark. From 1956 onwards a 'Royal Welsh' conference, held either in north or south Wales, was organized alternately with a tour to a foreign country, which until 1961 included Ireland and, on two occasions, Scotland. The annual spring tour to other countries continued until 1970.

Another innovation arising out of the working party's recommendation in 1953 was the institution in the mid-1950s of a Gold Medal Award in recognition of outstanding service to Welsh agriculture. Appropriately named the 'Bryner Jones Gold Medal of the Royal Welsh Agricultural Society', it was first presented at the December Council Meeting of 1957 to Emeritus Professor Thomas James Jenkin, sometime director of the Welsh Plant Breeding Station – founded in 1919 – and chairman of the National Institute of Agricultural Botany, Cambridge. As a farmer's son from Maenclochog, Pembrokeshire, Jenkin had received no formal secondary education before going to Aberystwyth as a student, but he became a scientist of great distinction and a pioneer in grassland research.

Other recommendations like the implementation of an essay competition, supporting the University Court in its decision to establish a veterinary school in Wales and the revival of a *Welsh Journal of Agriculture* were taken up by the Society, but to no avail. Whether successful or not in its pursuit of so varied a range of recommendations, the Society was clearly energetic in pursuing activities and supporting campaigns outside the show during these post-war years. Moreover, there were other out-of-show activities and competitions promoted by the Society alongside the aforementioned, some of which were a continuation from the pre-war years and others new to this period, activities which will now be briefly mentioned.

The quality of the Society's long-standing annual *Journal* was to improve markedly during the 1950s, a development commented upon by Professor R. G. White of Bangor in 1960. Although he had believed that the *Journal* had not been a publication worthy of the Society in the immediate post-war years, the excellent 1959 *Journal*, in his view, compared favourably with the publications of kindred societies. Much of this improvement stemmed from the efforts of a reinvigorated editorial committee – which occurred in the mid-1950s – under the chairmanship first of Major Gibson-Watt and, from 1956, of the Honourable Islwyn Davies, a staunch supporter of the value of the *Journal* in enhancing the Society's appeal. As editor from 1955, Dr Richard Phillips also played a key role in ensuring that each issue contained articles of special interest to Welsh farmers in both English and Welsh. Importantly, with the help of the Lady Roberts Trust Fund, the editorial committee was responsible for the periodical issue (two or three a year) from January 1962 of the *Royal Welsh Review*, which was designed to act as a link of information between the Society and its members.

Since pre-war days the Royal Welsh Agricultural Society had carried out secretarial duties on behalf of the Welsh breed societies, duties which increased appreciably over the course of the 1950s. This secretarial work, although of great importance to the breed societies, was a costly undertaking for the Society itself. Unfortunately, down to the mid-1950s the cost had not been fully covered by the breed societies' contributions and, considering the substantial services it rendered to them at its own expense, it was understandable that there was disappointment at the very low percentage of their members who also joined the Royal Welsh Agricultural Society. In 1956-7 an appeal was thus made to the societies for better support. By this time, secretarial responsibilities covered not only the Welsh Pony and Cob Society and the Welsh Sheep Flock Book Society but also the newly formed Welsh Halfbred Sheep Breeders' Association. A change in the method of grant aid by the first two of these breed societies was instituted from 1958. Henceforth they would contribute secretarial grants on the basis of 15 per cent of their turnover during the previous twelve months. However, even this did not sufficiently ease the financial burden incurred, a position made all the more unacceptable by the continuing reluctance of breed society members to join the Royal Welsh Agricultural Society. Accordingly, at the close of 1960, the Society agreed to invite the Welsh Pony and Cob Society and the Welsh Halfbred Sheep Breeders' Association to make contributions equal to 90 per cent of the direct costs of the salaries attributable to the work of the society concerned, a new arrangement which was accepted by the Welsh Pony and Cob Society by March 1961.

Apart from the show section, the Forestry Committee conducted annually from 1950 onwards a Woodland and Plantation Competition, which covered three or four counties each year in rotation. Judges prepared a detailed report for the landed estates and farmers, and a general report was distributed. Visits to the farm of the competition winner were organized. Inter-County Hill Flock Competitions were also sponsored by the Society from 1955 onwards. Later, in 1961, a Farm Machinery Maintenance Competition – the brainchild of W. J. Constable – was implemented. Presented by Sir Bryner's daughter in 1957, the Sir Bryner Jones Memorial Trophy competition was of considerable value to Welsh agriculture in general and it was agreed in 1959 that the trophy should figure among the Society's major awards. Whereas down to 1960 the competition was restricted in geographical area but open to many kinds of farm and farming practice, from 1961 it became open to the whole of Wales but restricted to one aspect of farming, namely, dairying in 1961 and beef production in 1962. Every endeavour was made to hold a field day on the farm of the winning entry. In similar manner, in 1960 the family of the late D. Walters Davies (past director of the National Agricultural Advisory Service, Trawsgoed) donated a trophy to the Society to perpetuate his name. The upshot was that, with the assistance of the Welsh Seed Growers Federation Ltd, a certified seed crop competition was organized annually.

Indicative of its standing, the Society forwarded views on procedures for combating foot and mouth disease to the Ministry of Agriculture. In November 1952 it submitted a solicited memorandum outlining its views on the Ministry's handling of the recent crisis which had mercifully barely touched Wales. The Society's chief veterinary officer, T. H. Jones, subsequently attended a committee meeting in London in January 1953 to support the views expressed in the memorandum and stated that his Society looked forward to the day when a swift diagnosis method would be perfected, thus making it possible to vaccinate all contact animals and thereby obviate the need for slaughtering them. He also informed the committee that one of the Royal Welsh Agricultural Society's criticisms of the Ministry's procedure related to the delay in advising the cancellation of the classes for cloven-hoofed stock at the Royal Welsh Show at Caernarfon in 1952. Furthermore, he personally believed that migratory birds were largely responsible for carrying foot and mouth infection from the Continent. Concerned at the extent to which Caernarfonshire hill farmers had been so seriously affected by the disease in 1957 that compulsory slaughter of complete hill flocks had been carried out, the Society set up an ad hoc committee in 1958 to investigate the special problems which arose whenever an outbreak occurred in a hill-farming area. Its report, completed by December, was sent on to the Ministry.

CHAPTER TEN

'An Act of Faith'

THE MOVE TO A PERMANENT SITE

Although a working party to examine the feasibility of a permanent site was not set up until 1958, suggestions that the Society should move to a permanent centre or semi-permanent sites were increasingly heard from the early 1950s. Indeed, the Monmouthshire county meeting had called for such a move as early as 1927! The Carmarthenshire county meeting began the snowball effect in 1950 by urging Council to consider the advisability of a permanent site, and the following year saw Breconshire push the selection of two or three semi-permanent sites in Wales for the purpose of holding the annual show and with these venues to be made available for experimental centres outside the period of the show. However, on the understanding that it was possible to secure sites for several years, the Society decided in autumn 1952 that the current policy of migratory shows should continue. 'M.R.' of the *Liverpool Daily Post* applauded the decision, believing that the very purpose and nature of the show, like that of the National Eisteddfod, would be defeated and damaged if such major national events ceased to be itinerant. Apart from espousing the virtue of the migratory principle under Welsh conditions, the Society recognized that the cost of buying a permanent site was beyond its financial resources.

The issue of the establishment of a permanent site was once again raised at the annual general meeting of members on the Machynlleth showground in 1954, in all likelihood out of frustration at the deplorable conditions underfoot at that ill-starred event occasioned by atrocious weather conditions. Once again, the financial implications of a permanent site persuaded the Society that it was impracticable to proceed with the proposal. County meetings, the democratic organs of the Society, nevertheless kept up the pressure on the Council in 1955. Cardiganshire's request that Council might consider the advisability of consulting the United Counties Agricultural Society with a view to jointly acquiring a site where the Royal Welsh Show could be held every five years or so was similarly matched by Carmarthenshire's proposal that the Council consider the selection of Carmarthen as a semi-permanent site. No doubt flushed by the marvellous success of its hosting the show in July 1955, shortly afterwards Haverfordwest Borough Council invited the Society to consider the racecourse in the town as a semi-permanent site for

44. Mr J. E. Gibby, president of the Society, at the 1966 show.

the show. In 1956 it was the north's turn to petition, with Denbigh-shire and Flintshire requesting a permanent site for the show. Once more the proposal was turned down on the ground that it was beyond the Society's means. Questions regarding the Society's policy on a permanent show site posed by members at the annual general meeting in July 1957 met with the same response that it was not in a position to undertake the very considerable outlay involved.

The dam of resistance was breached at the close of that year when, at its meeting on 20 December, the Council – in response to a notice of motion of J. Morgan Jones that the Society set up a working party to explore the question of a permanent site – instructed the Finance Committee to submit the names of persons to serve on a show site working party and to suggest terms of reference. As chairman of the Local Executive Committee for the 1957 show at Aberystwyth, Morgan Jones had come to harbour doubts about the Society's continued existence in the light of the fact that it had to rely so much on a local fund, for, with the choice of suitable sites gradually diminishing, there was a danger that the same people would be asked to support the local fund. He also felt it incumbent on the Society to improve its facilities to members and give much better service. After membership of the working party had been established, comprising twenty in all, at the first meeting on 5 June 1958 J. E. Gibby, of Pembroke, a joint vice-chairman of the Society, was appointed chairman. Its terms of reference were quoted as follows: 'To consider, in all aspects, the question whether the adoption of one or more permanent or semi-permanent sites would alleviate the financial position of the Society and report.' Any lingering doubts felt about the expediency of this course of action would have been dispelled by the muddy condition of the showground at the thirty-ninth annual show held in Bangor in July 1958. Under the heading 'One Field, One Site for the Royal Welsh?' a writer in *Y Cymro* for 31 July 1958 observed: 'Naturally the mud has greatly affected the size of the crowds, and when the terrible state of the field at Bangor was seen on the first day it was no wonder that there was great enthusiasm among the Royal Welsh Agricultural Society members for the principle of ensuring a permanent site for the great annual show.'

Over the prolonged period of discussion from 5 June 1958 to the Council's momentous decision on 16 December 1960 to purchase the permanent site at Llanelwedd, a great amount of work was carried out by the working party in particular, but also by the Finance and General Purposes Committee and the Honorary Director's Committee. Much information was placed before the working party, including at the outset four memoranda submitted in turn by Arthur George, J. Morgan Jones and Lieutenant Colonel R. E. B. Beaumont jointly, Alan Turnbull and Lieutenant Colonel G. E. FitzHugh, as well as details of actual financial costs incurred by other similar societies that had embarked on permanent site schemes. During the two-and-a-half years of its deliberations the working party reported progress to Council at intervals. Progress in reaching the big decision came slowly and haltingly, and naturally involved at times some division of opinion between various members of the working party. By the time of the Council meeting in June 1959, J. E. Gibby was in a position to announce that the establishment of a permanent site of bricks and mortar or concrete would be a very costly business and could not be undertaken for less than £250,000. This was thought to be out of the question and Lieutenant Colonel FitzHugh's scheme for a permanent show-ground of just twenty to twenty-five acres which would cater for a number of sectional shows during the year was, though attractive in some respects, deemed inappropriate. Instead, the working party had decided that the best course would be to erect buildings of a less permanent nature and more or less in keeping with the present structures whose cost could be contained within the region of £150,000. Mindful that one of the objects of opting for a permanent site was to reduce the heavy annual erection costs, the working party had agreed on the carrying out of six basic essentials over the first four years – purchase of the show site and parking areas; removal of hedges, levelling, draining and seeding; fencing showground perimeter; laying of water supply, electricity mains and permanent roadways – following which a programme of timber or semi-permanent buildings was to be drawn up.

In the course of reaching this decision to investigate further the possibilities of the development of a permanent site, different points of view, as already intimated, were expressed within the working party. Notably, Lieutenant Colonel FitzHugh and Alan Turnbull were opposed to the idea. Lieutenant Colonel FitzHugh's opposition was founded on his reckoning that it was financially impossible to stage a show of its present size and scope on a permanent site in Wales and also on his conviction that it would depart from the Society's aim of taking the show to the bulk of the Welsh people. Alan Turnbull likewise claimed that the farming community benefited most from the old migratory system. On the other hand, weighing the advantages and disadvantages of a permanent site in their joint memorandum, J. Morgan Jones and Lieutenant Colonel Beaumont pointed to the balance of advantage lying with the establishment of a permanent showground at this critical turning point in the Society's history. For them, the most telling arguments in favour were: the prohibitive costs of erecting showgrounds under the migratory regime; the precautions necessary to guard against the ill effects of wet weather at peripatetic shows had incurred heavy expenditure which would be significantly reduced on a permanent site with hardcore roads; tented accommodation, in addition to the possibility of wet weather conditions, worked against attracting the best caterers; under the old system of migratory shows, lavatory facilities and accommodation were unavoidably 'sordid and unpalatable'; year after year many people had gained entry to the show without paying, a practice impossible to prevent on temporary showgrounds lacking secure perimeters; the expanding size of agricultural shows generally meant that it was difficult to obtain suitable sites; the finding and preparation of a new site each year entailed a big effort on the part of the inhabitants concerned and imposed on them the burden of raising a local fund, on which the financial success of the show had come to depend more and more; and the improved facilities of a fixed showground would attract increased membership. These advantages, they maintained, outweighed the drawbacks of a permanent site, which included the loss of local interest and enthusiasm aroused each year in the show area, and the financial problem of acquiring and equipping a permanent site, which was admittedly 'tremendous' bearing in mind the Society's relatively low reserves and which had hitherto deterred it from contemplating a permanent site. If the balance was nevertheless in favour, Morgan Jones and Beaumont recognized that to change to a permanent site with its present small capital reserve and low membership would be 'an act of faith'. Therefore, at the very outset of taking this necessary step, the Society would have to make an all-out effort to attract the support of the Welsh farmer for the venture in the way of securing his membership and subscription to the capital fund which would have to be raised.

The means of winning the support of the Welsh farming community were deliberated upon at a meeting of the working party at the end of October 1958. It was believed that an effective way of assessing the level of support for the project would be for the secretary to address county branch meetings of the National Farmers' Union. That public support proved to be an awkward problem was reflected in the working party's press statement released early in 1960 and stressing the Council's firm conclusion reached at its previous December meeting that unless the Society had a regular membership of 10,000 – or about four times the present number – no proposal for a permanent site could be seriously entertained. To that end, the statement continued, the working party had been asked to prepare a scheme for the recruitment of the required membership and to demonstrate how a permanent site could be financed. Accordingly, the abiding problem of membership recruitment was fully discussed at a joint meeting of the Finance Committee and Working Party on 4 March 1960, and centred on a memorandum prepared by Moses

Griffith which argued that the Society was not keeping pace with important developments and problems directly affecting Welsh farmers. After various views were expressed regarding the best way to boost recruitment, the crucial decision was taken that the Council be informed that in the opinion of the joint meeting the establishment of a permanent site for the show was one important factor towards increasing the Society's membership. The Council's position that a total of 10,000 members was necessary before the project could be launched had been turned on its head! The momentum for a permanent site was seemingly unstoppable, particularly in view of Council's decision on 18 March 1960 that the establishment of a permanent site was one important means of recruiting new members.

The vital step in crossing the Rubicon was taken on 23 May 1960, when the working party unanimously resolved 'that the Council be recommended, in view of the Working Party's firm conviction that it would become more and more difficult to achieve local fund targets for peripatetic shows, the time has now arrived when the Society must acquire a permanent site for the Royal Welsh Show', and that 'Council be recommended to consider acquiring a suitable permanent site in Aberystwyth'. There is no doubt about the importance of this meeting as a watershed, for a firm statement of policy had now been made. No doubt helping the working party reach this decision was Alan Turnbull's statement that while he had never really favoured a permanent site project he was being forced to change his mind when confronted with the increasing difficulties in raising local funds. In agreement with his views were Lieutenant Colonel Beaumont and also J. Morgan Jones; the last mentioned illuminated the Society's position by claiming that a peripatetic show, depending as it did on a local fund and on an ever decreasing number of suitable sites, was 'slowly grinding to a halt'. In a measured statement, Morgan Jones observed that whilst it was appreciated that local funds had, with few exceptions, exceeded their target, the drain on the resources had been such that it would make things extremely difficult if the show were to visit those same counties within the near future.

Council met on 10 June 1960 and by an overwhelming majority of 32 votes to 3 adopted the working party's recommendations. Major difficulties remained to be faced, however, above all the question of the capital necessary to purchase, establish and develop the site. Once more the working party was entrusted by Council with two new tasks, namely, that, calling on expert advice, they investigate the whole subject of financing a permanent site and that all Welsh farmers, industrialists, Society members and other involved persons be asked whether they would lend their full support towards establishing a permanent site at Aberystwyth.

By November 1960 the working party – guided in its decisions by the three

45. The old Llanelwedd Hall, before the fire of 1951.

chosen investigators of the merits of the various sites proposed, namely, Alan Turnbull, J. E. Gibby and R. L. Jones of Newcastle Emlyn – believed that Aberystwyth was unsuitable as a permanent venue, not least because only seventy acres were available to accommodate a show which required a hundred acres. Other possible suitable sites at Builth Wells, Machynlleth, Llandinam and Caersws were discussed, the last of which, members heard, might seemingly be developed into an 'overspill town' of Birmingham. Furthermore, dual sites, one in the north and the other in the south, were contemplated, only to be rejected on account of double running costs. Having decided that difficulties would arise in the case of purchasing sites at Machynlleth, Llandinam or Caersws, the only positive option open to the Society was the Llanelwedd Hall site near Builth Wells. Legal problems might arise over the Machynlleth site since it was common land, while Builth, with Llandrindod Wells nearby, was more attractive than both Llandinam and Caersws from the point of view of available accommodation. The final push towards clinching the decision over the permanent site in favour of Builth occurred at a meeting of a sub-committee, comprising J. E. Gibby, Lieutenant Colonel FitzHugh, Dr T. L. Davies and Arthur George, with locally based

Alderman Harold Edwards and his colleagues, at Builth on 10 December 1960. The upshot was that Lieutenant Colonel FitzHugh reported to the Council a week later that the 176 acres – 104 of which were flat, the rest on a slope – would have to be purchased and, provided it could be bought at market value, it would definitely meet the Society's needs. Council thereupon bravely resolved that the Society, through its Finance Committee, should negotiate for the purchase of the site.

At the Finance Committee meeting on 12 January 1961 its chairman, Lieutenant Colonel FitzHugh, suggested that, apart from any negotiations to purchase the Llanelwedd site, the Society should engage Harold Edwards to prepare a valuation of the land. Such an announcement saw the temperature rise; a forthright Moses Griffith claimed that he was very unhappy about the way in which a discussion of such magnitude had been taken so suddenly and without sufficient information being given to the Finance Committee and Council members. Similarly, R. W. Griffiths expressed concern at the apparent uncertainty of members as a whole and argued that if a permanent site project was not adopted with full confidence it would be better not to pursue it at all. He wondered whether Builth Wells was the right area. When put to the vote, with Moses Griffith the only dissentient, it was resolved that Alderman Edwards be instructed to prepare a professional valuation of the land and that negotiations for its purchase be pursued. Meeting again on 9 March, Harold Edwards reported that such was the local enthusiasm that a fund-raising committee had been set up which had already pledged £10,000 towards the site's purchase. Moses Griffith once again registered his unhappiness with the position and asked that it be recorded that he had abstained from voting on this matter because he considered that a permanent site at Builth Wells would not serve the whole of Wales. The meeting recommended that Council purchase Llanelwedd for a sum not exceeding £32,000, including tenancy compensation. By virtue of the good offices of the Builth Wells Permanent Site Committee, chaired by Alderman Edwards, nearly £22,000 was raised, and a nest egg of over £15,000 from the surplus of the Gelli-aur show meant that the financial wherewithal was available to meet the total cost of £38,497 10s. required for the land, compensation moneys to the tenant, and some legal costs, without having to realize any of the Society's investments. This – the most important project the Society had ever undertaken – had reached completion by summer 1961 with full possession of the site following in May 1962. In the light of the prompt action of the Builth fund-raising committee it is clear that the Society had been presented with a fait accompli and had little choice but to accept. (The situation thus bore resemblance to the choice of Aberystwyth as the first centre for the show in 1904 when a hefty donation and a suitable site were both on offer.) But the decision was not reached without misgivings. Alan Turnbull was later to recall to David Lloyd of the *Liverpool Daily Post* that the Society was nervous about the poor support given the earlier show held at Llanelwedd in 1951 and that many felt it was too close to the English border and those from the north thought it was too far south.

Taking the decision, however, did not signal that it was now time to relax. Indeed, in October 1961, Lieutenant Colonel FitzHugh reminded Council of the very important step which had now been taken and the grave responsibilities attached to the decision. The Society could not hope to lay out a showground without first of all deciding on a policy for the site's use. Warming to his role as a revered counsellor, FitzHugh stressed that there were very serious questions to be settled, questions which Council ought to be in a position to answer by early next year. Council duly resolved that the management of the site be entrusted to the Honorary Director's Committee. We shall return to the way in which the Society faced up to the new situation and settled down after its nomadic lifestyle in a later chapter.

All that remains to be said at this stage is that it was only to be expected that the decision would meet with disapproval as well as commendation among the wider membership and public. Moses Griffith's belief that Llanelwedd was an inappropriate centre for the general body of Welsh agriculturalists was shared by the Welsh Black Cattle Society which wrote to the Royal Welsh Agricultural Society in May 1961 protesting against its decision and asking Council to reconsider on the grounds that the permanent site of the Royal Welsh should be accessible to the majority of Welsh farmers. Roland Brooks, agricultural commentator of the *Western Mail* and himself supportive of the decision to settle at Llanelwedd, reported in July 1961 that Welshmen in north Wales had complained that the site at Builth was too distant, had talked of starting a 'splinter' show, and that there had been murmurings from other parts of Wales.

It could not be fairly said that the Society had rushed its decision. Furthermore, by choosing a permanent site it was following the trend of most of Britain's leading agricultural societies. Obviously some of the members harboured misgivings throughout, as indeed did Lieutenant Colonel FitzHugh, who nevertheless played such a vital guiding role in the whole process of transition that J. E. Gibby, vice-president of the Society and chairman of the working party, generously described him as the 'Architect of the Modern Royal Welsh'. From the early 1950s he had opposed a single site but, as a loyal officer of the Society, he complied with its wish once the nearly unanimous decision had been taken and declared his fullest support in establishing the Society's headquarters and annual show at Llanelwedd. His advice to Council members on 29 March 1962 was couched in allegorical vein

46. Lieutenant Colonel G. E. FitzHugh (second from the right) in consultation with members of the Society and others over preparatory showground operations on the permanent site.

in viewing the move to a permanent site as the equivalent of entering a state of 'matrimony' from the 'bachelor' existence previously led by the Society. He confessed himself rather concerned about the fate of 'bachelor' friends. Whilst one body of opinion considered that it would have been better for the Society not to have 'married' at all, a second body believed that the Society had 'married' the 'wrong girl'. In his view, this sort of attitude never helped to make a success of any marriage. But the crucial question was: did the two sections feel so strongly in their viewpoint that they were unable to support the venture. Whilst there were certainly those who disagreed with this new important step, it was clear that they *accepted* the position. It was imperative, he urged, that such acceptance should be complete, and that the Society should be single-minded and tenacious in seeing the venture through.

PART FOUR

From Crisis to Triumph

THE LLANELWEDD YEARS, 1963–2004

CHAPTER ELEVEN

The Llanelwedd Years, 1963–1975

GLOOM AND DESPAIR

The abandonment of the moving show in favour of a permanent site – an initiative which was in line with the practice of the Royal Agricultural Society of England, which moved to Stoneleigh (1963), the Royal Highland to Edinburgh (1960), the Bath and West to Shepton Mallet (1965) and the Three Counties which moved to Malvern (1958), all in response to the very high cost of staging a peripatetic show – meant, of course, that ever tighter links had to be forged with the localities. Towards this end, we have noted earlier, in 1962 the Society incorporated county advisory committees into its constitution, the nucleus of which were composed of Council members resident in each county and with each county served by honorary county secretaries. At the same time the work of developing the Llanelwedd site was placed in the hands of the Honorary Director's Committee. Lieutenant Colonel FitzHugh and his support team faced a considerable, not to say daunting, challenge.

The essence of the problem lay in the Society's continuing slender financial resources. Its officials were anxiously aware in early 1963 that a total membership of 3,649 was pathetically small and that as many as 10,000 subscribers would be necessary to ensure the future success of the new project at Llanelwedd. Disturbing, too, was the Finance Committee report presented to Council in June 1963, which showed that one of the arguments for a permanent showground freely used in the past was a fallacy: the old assumption that by moving to a permanent site a large part of the annual cost of erecting the show would be saved was only true up to a point; constructing roads, for example, would be a saving, but in many cases the maintenance and depreciation of the kind of buildings that were envisaged would exceed the hiring charge of the tents and shedding they replaced. With allowance for continued inflation, the report concluded that the Society's annual expenditure was likely to increase rather than decrease.

A careful estimate of the costs facing the Society over the next ten years or so was also noted in the report. The first phase included the purchase of the site, the compensation to the outgoing tenants and other essential work necessary for the holding of a show on the ground in July 1963 – a total cost of some £67,000. In the second phase the renovation of the farm cottage (£500) and provision for staff housing (£14,000) should be dealt with before the 1964 show if sufficient sums were raised to allow it; on top of this, the second phase included facilities for stockmen (£9,500), additional services and a provisional figure of £5,000 for the purchase of second-hand caravans as accommodation for show officials. It was calculated that, in all, £33,000 would be required to cover the costs of phase two. Developments in phase three, whose total cost was estimated at £70,000, would include completion of services, a sewerage scheme (£13,000), a proportion of toilet and associated facilities (£16,500), further provision for stockmen (£4,000), provision for trade exhibitors (£5,000), and laying of roads (£25,000). In the somewhat distant future of 1970–1, phases four and five would be undertaken, their developments covering an extension of the Council buildings, toilet and associated facilities, additional cubicles for herdsmen, concrete flooring for livestock accommodation, and a permanent grandstand. Approximately £80,000 was required to complete these last two phases. All five phases taken together meant that the total capital costs, including the purchase of the site, was in the region of £250,000, the figure previously envisaged by the Council.

For its part, the sub-group entrusted with the difficult task of making a projection of five years' income reported that the total annual income needed to be increased from the £55,000 estimated in 1963 to £80,000 over the next five or six years. Recognizing that this was a very ambitious target, they maintained that increased income should be sought from membership, gate receipts and pre-show tickets, as well as from non-show activities, and they urged the necessity of launching an intensive membership campaign at an early date. On the basis of this information, the Society initiated two very important appeals in 1963 and 1964. First came the National Appeal Fund, launched by Viscount Emlyn at Llandrindod Wells on 28 November 1963, with the objective of providing a capital fund of £250,000 for the purpose of developing the Llanelwedd site as the permanent headquarters for the Society and its show, and as the farming and rural life centre of Wales. Sitting as members on the Appeal Committee under Lord Emlyn's chairmanship were the Honourable Islwyn Davies, A. B. Turnbull, Sir David M. Evans-Bevan, Martyn Evans-Bevan and Bevington R. Gibbins. This committee worked in the initial stages with the campaign director provided by the Wells Organisation, a professional external body hired by the Society to give advice regarding the best way of organizing the appeal.

In early 1964 came the Society's appeal for a 100 per cent increase in membership, which was viewed as essential for its very survival. Matters had been made all the more pressing by the considerable financial loss sustained by the first show held at Llanelwedd in July 1963, mainly because of a low gate occasioned by the delayed harvest. It was also the case that under the old system of mobile shows the local fund appeal always boosted membership in that county by up to a thousand or more, many of them, unfortunately, for that year only. It was all the more important, therefore, that the people who had hitherto supported 'occasionally' under the old type of local fund would now change to an annual membership. The membership launch was held on 2 March, before which date county meetings were held in the thirteen Welsh counties and knowledgeable subscription collectors nominated for each district. Clearly the Society's officials believed that the main impetus for a membership drive must come from the county advisory committees and these same officials were keenly aware that the response to both appeals would determine the rate at which the Llanelwedd site could be developed. Already by the beginning of 1964 there was a considerable demand from existing members for the provision of better facilities at the show.

Besides drawing up plans for the site development and launching two appeals, important constitutional, administrative and publicity changes would have to be made if the Society were to meet the challenges posed by the move to the permanent site. Before that move, as we have seen, the Society was not a corporate body and its government was vested in the Council. Within broad limits, and with some occasional reorganization of the committee structure to meet changing needs, this structure had served its purpose. However, the goal of acquiring charitable status was adopted at a Council meeting at Dolgellau in summer 1965, and the Society realized that the move to Llanelwedd meant that it was growing into something much greater than a three-day show. In the following November Council decided to register the Society as a charitable limited company so that it could more effectively perform its varied functions and develop Llanelwedd on sound business lines as the main centre of agricultural activities in Wales. As a substantial property owner which intended to raise large capital sums, it was essential that the Society should become a corporate body with limited liability; otherwise each individual member could be held personally liable for any of the Society's debts.

In view of the great amount of extra work to be done in developing Llanelwedd as an all-the-year-round demonstration centre, a satisfactory administrative set up was also required and towards this end in 1965 the Society underwent a streamlining exercise whereby the executive control of its affairs was entrusted to a much smaller committee. It was felt that the cumbersome Council of more than 150 members, meeting only three or four times a year, was inadequate to deal with urgent matters. But while a small executive, styled the Board of Management, would henceforth run the charitable limited company, power would still be firmly held by the great mass of members working through the all-important county advisory committees. The latter would, in turn, elect Council members from within their own ranks. The basis of representation at grass roots level was that of six members on each county advisory committee for every fifty Society members (or part of fifty): thus, the more members within the county concerned, the greater the strength of the county committee. Council members, for their part, would consist of three members for the first hundred members (or part of hundred) of each advisory committee with one added for every extra hundred committee members (or part of hundred). In so far as the Board of Management was concerned, meeting once a month its membership was to be composed of officers of the Society and one member per county elected by the Council. Every county having more than 1,000 members would have an extra member. Both the Council and the Board of Management were granted powers of limited co-option. Although occasional criticisms were to be heard of this new constitution, Lieutenant Colonel FitzHugh, once again the architect of this new form of government, was at pains to point out that it was as democratic as anything that could be humanly devised and that the members in fact had been given extended powers. Delayed in its implementation, the Royal Welsh Agriculture Society Limited became effective in January 1967.

The Society's officials were also aware of the need for an improvement in publicity – far beyond its level in the migratory years – in order to promote the new venture in its manifold activities. Effective publicity would necessitate advertising the Society in those circles from which it was looking for the most substantial donations to the capital appeal, namely, industry and commerce, publicizing the membership drive, and promoting the show as well as the many all-the-year activities of the Society. After settling at Llanelwedd it was obviously vital for the Society to maintain contact with those counties remote from the permanent site. With regard to the show itself, the organizers had to face the fact that they had lost the publicity which they used to generate in the process of raising the local fund.

Following the disastrous attendance of 42,427 at the first show at Llanelwedd in 1963 – John Wigley recalled that 'it was with a sinking feeling that one walked up and down the empty avenues on the third day' – the decision was made to spend more money on publicity specifically directed at the urban population of the Rhondda valleys, and with the 65,000-attendance in July the policy seemed to have paid off. Mid-1964 saw the setting up of a Publicity and Publications Committee, on which experienced journalists, Sylvan Howell and Roscoe Howells, would play an important public relations role. Howell became public relations officer of the Society in autumn 1966 at the same time as he undertook the editorship of the *Journal*. Yet, according to Lieutenant Colonel FitzHugh, in his address to the Board of Management in November 1967, the state of the Society's public relations was still unsatisfactory: few of the Society's officials really understood it and many were loath to vote the amount of money which the Publicity and Publications

Committee demanded. Prompted by J. Llefelys Davies, chairman of the committee, an effort was made to bring greater efficiency to publicity in early 1968, the outcome of which was the establishment at the end of the year of the Editorial and Publicity Committee composed of Richard Bowering (chairman), Alderman Harold Edwards, Sylvan Howell, Dr Richard Phillips, Elwyn Thomas, Mansel Davies, Llywelyn Phillips, David Lloyd and M. W. Jude.

A further publicity drive, including car and other competitions, occurred upon the appointment of Messrs Skinner and Co. Ltd, consultants, London, to fill the role of part-time publicity officer from April 1969 onwards. Unavailing endeavours were made by the company to attract Mary Hopkin, Harry Secombe, Richard Burton, Elizabeth Taylor and Tom Jones as visiting celebrities to the showground, as, also, to persuade the BBC programme, *The Archers*, to give the Royal Welsh Show a plug on its programme. For all the activities mounted as part of the new publicity drive, there was nevertheless some disquiet expressed at the want of press publicity in spring and early summer 1970 connected with the Society and the show. The situation was saved by John Kendall's help at short notice before the 1970 show; an agricultural correspondent for the *Farmer and Stockbreeder*, he was appointed as the Society's press relations officer in June 1971 and his contribution towards facilitating the continually improving public attitude and reaction to the Society and its shows, particularly evidenced during 1972, was acknowledged by the Editorial and Publicity Committee in early 1973. He would continue to play a key role in handling the Society's public relations up until the present day. From the end of the 1960s not just the agricultural press but the non-agricultural public, too, were targeted as can be seen in the publicity gained in newspapers like the *Sunday Mercury* and the *Birmingham Evening Post*. On an individual level, Charles Quant of the *Liverpool Daily Post* and Roland Brooks of the *Western Mail* were staunch in their support in these difficult years down to the mid-1970s. In 1975 a vital change occurred in the advertising for the show – long entrusted by the Society to its publicity agents, Messrs Creighton Griffiths (Advertising) Limited – for this was the first year that television was used; indeed, television advertising was the main media utilized that year, thanks to the cooperation of HTV which allowed the Society to use parts of one of its films at no charge.

One further development which rendered the Society better equipped to cope with the responsibilities arising from the move to Llanelwedd came in a reorganization of administrative methods and staff. It was envisaged that from 1964 Arthur George would assume greater managerial responsibility with the intention that this would relieve committees of much of the detailed work which was tending to arrest progress. Accordingly, in 1965 he became secretary-manager, and John Wigley was entitled administrative secretary and finance officer. Despite some improvement, the administration under the new constitution continued to be overburdened with committees, and in November 1967 the new Board of Management appointed an Executive Committee in place of the Estate Management and Financial Control Committees. The Executive Committee consisted of officers of the Board of Management and seven other members. It was to meet once monthly to exercise financial control and deal with issues other than policy matters and it was agreed that this committee should not allow itself to become involved in administrative matters which ought to be dealt with by the staff. The other committees were instructed to reduce their committee work by meeting once annually, while the Board itself decided to meet every other month. Further reorganization followed Lieutenant Colonel FitzHugh's address to the Society in October 1972, wherein he advocated a new committee charged with recommending long-term policies to the Board of Management. The resulting Policy Committee – in effect a small body of directors – met for the first time in April 1973 and comprised the chairmen of Council, the Board of Management and the Finance and General Purposes Committee, along with two members of the Board, namely Val Morris (Breconshire) and Tudor Davies (Glamorgan).

Alongside this overhaul of committee procedure, the Board of Management in November 1967 considered what staff would be required to enable the work of the Society and of the new commercial company set up in August 1966, Royal Welsh Agricultural Society Enterprises Ltd, to be carried out. In addition to Arthur George and John Wigley, a new development officer was appointed to supervise site development and to organize the affairs of the commercial company, tasks requiring a particular kind of professional competency. There were difficulties in recruiting a suitable candidate for this demanding post, but in late 1968 and 1969 the Society benefited from the part-time assistance of R. J. Morris, the retiring surveyor for the Llandrindod Town Council, who helped with the preparation of plans and kept an eye on building work in progress. August 1973 saw a big change in administration upon the retirement of Arthur George, who had been at the administrative helm since 1948. The equally long-serving deputy secretary, John Wigley, was promoted to secretary-accountant, and a new post of chief executive was filled by 39-year-old Cambridge graduate and an executive of Britannia Airways Limited, Philip Phillips. Commencing duties in February 1974 Phillips's flair was to leave an indelible stamp on the Society before his departure in March 1975 for a managerial position in New Zealand. Wigley was later promoted to secretary-manager in June 1975.

One further administrative innovation following the move to Llanelwedd was the greater emphasis on the county unit and the county advisory committee which, as we have seen, with the new constitution forged in 1965 was to be the first stage in the Society's democratic organization. The maintenance of the all-important link between the Society and the counties was seen to be a crucial function of the Council. At last the Society seemed to have understood fully that if this relationship were not nurtured carefully then membership would fall away and show attendance would also suffer. Nevertheless, in early 1971, members of the Board of Management were concerned at the waste of time, money and effort spent in the current system of county advisory committees, which were generally poorly attended. Accordingly, worried about the communications gap between the centre and the counties, and aware of the unsatisfactory county secretarial set-up in 1971, the Society, then in a critical financial situation, established two regional organizers, for the north and south respectively, who, besides recruiting members and helping to establish 200 Clubs, would assist in servicing county advisory committees.

If these changes were introduced to equip the Society to cope with the new challenges, the smoothness of the transition depended a great deal on the success of the Society's attempts to raise capital funds and membership numbers. As far as the National Appeal Fund was concerned, the pattern of response was not as the Society would have liked, and by spring 1965 officials were much concerned that capital expenditure was running ahead of receipts from the Appeal Fund. Moreover, they were conscious of the fact that only a limited sum for expenditure each year would be available under the system of covenants introduced in 1964. Thus, even though the Fund total was approaching £140,000 by the financial year ending 31 March 1965, this was an amount spread over seven years. It was obvious to them that under such a system the Society would be overspent on capital for some years. In the early months of 1965 the Appeal Fund had been assisted by the county advisory committees who organized collections within their own counties. Influenced by Pembrokeshire's success in achieving its self-imposed target of £12,000 for the National Appeal Fund in 1966, the Society decided that in future the National Appeal Committee should concentrate its efforts on a specific or 'feature' county each year. Cardiganshire thus undertook to raise £20,000 in 1967. Spearheaded by the Society's president, Dr Jenkin Alban Davies of Brynawelon, Llanrhystud, by May 1967 the National Appeal Fund had risen to £178,600, with £10,000 collected by covenants from Cardiganshire since February. The lag of National Appeal Fund receipts behind the development of the showground was to be an ongoing problem, however; for the year ending 31 December 1967 the development of the showground was thus ahead of revenue received from the fund by approximately £35,000 – equivalent to the overdraft on the general account. Partly because the target of

£250,000 set at the outset was not achieved, the programme of development planned for the years 1962–72, while executed in the first three phases, was not accomplished for phases four and five. At the end of 1971 it was urged that the importance of a capital fund being raised under the feature-county system should be emphasized, since without a continued programme of appeal, development could not be realized. Inevitably there would be a considerable fall off in the National Appeal Fund after 1973, for this was the last year of the covenants inaugurated during the Cardiganshire drive in 1966–7. Consideration was accordingly given to the advisability of launching another National Appeal Fund, but the Finance Committee in January 1973 felt it would be better to seek continuing support from the present donors and to request the assistance of the county advisory committees to effect this. Council was reminded by the Honourable Islwyn Davies in July 1973 that, quite apart from the generous contributions of the counties to date, the programme of showground development depended on the continuation of the National Appeal Fund for capital development; and he thanked those who had renewed their covenants to the Society.

Failure to reach the target set in the National Appeal Fund, combined with the fact that the cost of providing the capital facilities increased at a much faster rate than originally forecast, dictated that the speed with which the process of establishing the permanent site was implemented was much slower before 1971–2 than planned. This in turn meant that the erection cost of the show as itemized in the annual contractor's account was not reduced quickly enough and so remained a grave obstacle to the Society's attempt to achieve a surplus on the show account. Thus the Show Administration Committee of October 1972, in discussing the erection of new buildings for stock, agreed that it was essential 'if the Society was to survive' to erect buildings, a decision which would reduce the annual contractor's costs.

The decision taken in January 1973 not to launch another national appeal was due to the Society's placing greater priority on the membership drive which was so crucial in boosting its annual income and which by this time was in the doldrums. Under the chairmanship of J. E. Gibby, the Membership Drive Sub-Committee made excellent progress in its recruitment drive in every county in the years immediately following 1963, so that by 30 September 1964 there had been an almost 70 per cent increase in numbers since 31 October 1963, from 4,566 to 7,753. This campaign had been the most successful in the history of the Society and Edward Gibby referred in July 1964 to those who had been outstandingly successful in the enrolment of members, namely James (Jimmy) Prytherch (Breconshire), T. H. Jones and Ll. Thomas (Carmarthenshire) and William Evans (Monmouthshire). The momentum was sustained, and membership peaked around 10,000 in late 1967. There was a snag, however, for the same meeting heard from the honorary joint treasurer, L. Smith Davies, that members were 'not playing the game': the £2-privilege badge was being abused, and dishonesty was even greater under the joint membership of £3. It was generally agreed that there was a need to revise the membership privileges linked with an increase in subscription. In the teeth of much opposition from county advisory committees, membership subscription was raised in February 1968 to £3 and, for family relations, to £1 10s. Predictably, amendment of bankers' orders at the increased rate was a slow process.

Mainly because of this increase in subscription fees, by March 1970 membership had fallen to 7,000, of which, worryingly, 58 per cent were over the age of 55 and only 10 per cent under 25. By February 1971 a meeting of the County Organising Committee expressed concern at falling membership, which by this time was down to around 6,000. The financial position was so parlous – largely because show attendances were dipping well below the required 75,000 to 80,000 needed to produce a worthwhile result – that greater income had to be found immediately. Accordingly, the membership drive was renewed in 1971 and, as mentioned earlier, two regional organizers were appointed to recruit members, with a target of doubling the membership within three years. It was hoped that this, together with the establishment of twelve 200 Clubs, would secure revenue to ensure the Society's future. The situation was critical: Austin Jenkins darkly reminded the Financial and General Purposes Committee in April 1972 that, unless the Society made every effort to increase membership support by at least 1,000 a year, 'there was little hope of an improvement in the overall financial position'. As an inevitable result of the inflation then taking place, membership subscription was increased to £4 and family membership to £6 towards the end of 1972. An inspection of the draft accounts for the year ending 31 December 1972 made for grim reading, and so once again it was drummed into the Society by its chartered accountant, Charles Elphick, in February 1973: 'The Society must look to ways of increasing its income as a matter of priority and membership subscriptions appear to offer the best source.'

Frustratingly, no great headway was being made, however, in the renewed membership drive. Thus, total membership at 1 August 1973 stood at just 6,519, with well over 2,000 of this number having failed to convert their membership to the new rate of subscription. Inevitably, questions were asked about the effectiveness of the regional organizers, and the feeling was expressed that they were spending too much time on other events in the counties at the expense of work in connection with membership recruitment. At the beginning of 1975 the Society decided that the arrangement of regional organizers had not worked and thus abandoned it. Once again, faced with no choice in these days of inflation, the Society raised membership subscription rates early in 1975, with single membership increasing to £6 and family membership to £10.

Apart from income from membership, the other main source of revenue was, of course, gate receipts from the annual show. Unfortunately, attendances over the first four years, from 1963 to 1966, suggested that sufficient people could not be attracted to the show. It was also true that as membership increased, the paid gate decreased. Given that the costs of putting on the event were rising – the 1965 show cost £75,000 to stage, which was £18,000 more than the previous year – attendance levels were simply too low to enable the show to make a profit. In the face of the staggering increase in the cost of staging the 1965 show it was estimated that an attendance of 85,000 was needed for the Society to show a profit, and when rain caused a low attendance of 59,419 the loss of £10,445 was an inevitable outcome. Average attendance at the four shows between 1963 and 1966 was just 58,000, while the average loss amounted to £11,610. These losses were greater than the Society had feared and threatened to curtail its activities altogether. For all the hope expressed that 1967 – the first year since 1961 that the accounts showed a surplus, albeit a small one – marked a turning point in the Society's fortunes, this did not turn out to be the case. The familiar story of loss returned in 1968 largely due to low attendance figures at the show. No sudden lifting of the clouds was to occur, for the Society would face for some years to come the burden of an insufficient annual income further bedevilled by rising costs, particularly on showground erections expenditure, which proved very difficult to contain. Nothwithstanding the much-enhanced publicity for the show, the gate attendance stubbornly refused to approach the required 80,000 mark until 1973 when it rose to 77,024. A Board of Management meeting for September 1969 gloomily concluded that 'the difficulty appears to be that the public do not treat the Royal Welsh Show as the National Show and do not give it national support'. It never had done so, we must ruefully acknowledge. Despite good weather and a royal visit from Prince Charles as part of his tour following the investiture at Caernarfon Castle on 1 July, the 1969 show attracted only 72,840. Moreover, as already noted, the costs of showground erections aggravated the situation. In an earlier chapter we learned that the cost of staging the show annually had increased from £15,271 in 1947 to £57,832 in 1962. The chief reason behind the decision to move to Llanelwedd had been based on the belief that the rapidly rising costs of staging the annual show could be stemmed by the provision of permanent features and facilities, yet the show organizers were to be disappointed in their hopes of achieving big savings. An estimate of the cost of staging the 1969 show had it been erected on a peripatetic basis revealed that the show erection account would have been

£33,500 as opposed to £27,400 on the permanent site; as we have earlier observed, the speed of establishing the Llanelwedd showground was slower than envisaged, with the result that the erection cost of the annual contractor's account was never reduced below £27,000. Here we have to appreciate that the increase in labour charges between 1962 and 1970 was 40 per cent (14 per cent alone in the year 1969–70) and only as a result of close scrutiny on the part of those Society committees involved was the figure contained.

Low gates and rising costs meant that between 1963 and 1970 the show had made annual losses, except for that of 1969 which made a profit of just under £1,000. It had been clear to officials since March 1966 that sources of revenue apart from the annual show would have to be found, and in February 1971 the absolute necessity for a greater income immediately in order to ensure the financial viability of the Society was being emphasized. Such a serious financial situation led to the appointment of the aforementioned regional organizers, who were in effect salesmen representatives charged with recruiting members and with organizing 200 Clubs.

From 1966, following the loss-making shows on the new site, it was hoped that 200 Clubs and the activity of the new commercial company, the Royal Welsh Agricultural Society Enterprises Ltd, would provide the required additional revenue. The commercial company would be responsible for organizing events on the showground during the non-show periods of the year. Only in this way would the Society, as its first priority, be able to balance the income and expenditure of the year. Until this balance was achieved, the whole future of the Society would remain in jeopardy. The new company would protect the charitable status of the Royal Welsh Agricultural Society in so far as it would organize money-raising events other than those permitted within the charitable objects of the Royal Welsh Agricultural Society.

On the whole, this expectation of raising extra revenue was thwarted. A scheme of weekly contributions of 5s. in order to establish money-raising clubs to be known as a 'Royal Welsh 200 Club' in each county was suggested to the Board of Management in May 1966 by T. H. Jones of Llandeilo, honorary veterinary officer over the previous two decades. The scheme involved, subject to membership of 200 persons, a total contribution of £2,600. Prize money would amount to £1,450 and, on the assumption that the club's organization and expenses would involve another £150, this would leave a balance of £1,000 which would be handed over to the Society's current account. A recommendation was made by the Board that each county committee adopt the scheme, the public to be reminded that 5s. a week was just the equivalent of a packet of cigarettes and that a club member would have a sporting chance of winning a brand new car or £500. Although the J. A. George Club, based at headquarters, and the Breconshire Club – under the drive of Val Morris – were up and running by spring 1967, the response elsewhere was frustratingly slow for a Board of Management which saw the balance accruing from the clubs as a vital source of revenue and a quick way to bridge the gap in the yearly account in the immediate future. Significantly, the small credit balance of £366 in the 1967 account had been achieved through the receipt of £2,000 from the two 200 Clubs. Despite constant urging on the part of the Society's officials that they be set up, by September 1970 just 3½ Clubs were operating, namely, the J. A. George Club, the Breconshire Club, the Montgomeryshire Club and, though on a reduced scale, the Pembroke-shire Club. No real progress was made either under the regional organizers' drive between 1971 and 1974, but eventually clubs were also set up in Clwyd and Cardigan.

Nor did the raising of funds by means of sporting and entertainment events organized under the auspices of the Royal Welsh Agricultural Society Enterprises Ltd do much to swell the Society's needy coffers. Incorporated on 8 August 1966, the commercial company's two original directors were the Rt. Hon. Viscount Emlyn and the Hon. Islwyn Davies. Their initial task was to recruit support from outside commercial bodies for the company, whose success would be entirely dependent upon this. Reporting as chairman of the Board of Management to Council in October 1967, Lieutenant Colonel FitzHugh remarked that the commercial company had not acquired any momentum and therefore certain steps had had to be taken to bridge the gap by holding a number of small events on the showground, such as pony trotting, held on 23 August and 20 September, which had brought in receipts of £204. At the same meeting Alan Turnbull, the honorary director, suggested that motor racing should play a bigger part. In typically dynamic fashion, he strove hard as a member of the Motor Racing Circuit Committee to make this dream come true – he believed that the project would yield a minimum annual income of £5,000, almost matching the £6,000 rental that came to the Royal Highland Society from its circuit. For his part, Arthur George observed in March 1970 how it was hoped that this road racing circuit venture would realize such an income that the Society would be able to assist voluntary bodies, such as the Young Farmers' and Pony Clubs. Efforts to prepare plans for such a circuit over a period of some three years were to come to nought, however, because of lack of capital to fund the £70,000 scheme. It was simply not viable, given the prevailing economic climate of the country in general and of the car and ancillary industries in particular.

When the commercial company was set up, it had envisaged raising revenue to the tune of £5,000 or more a year. In reality, nothing near this was achieved and progress was slow and disappointing. Facing financial crisis in March 1971, the Board of Management was informed by its chairman, the

Honourable Islwyn Davies, that the various projects that were now in hand annually on the showground – including stallion shows, sheep sales, pony club rallies, Young Farmers' rallies, hound shows, autocross events, motorcycle scrambles and pony trotting – did not add up to an annual revenue of more than £1,000 to £2,000, whilst the gap that had to be bridged was a minimum of £10,000. That additional revenue so vitally needed was looked for from increased membership and 200 Clubs.

Almost miraculously, the gloom of 1971 gave way to quiet optimism by early 1973. Considerable pleasure came from the realization that 1972 was not only a year of consolidation but one of real progress. The Society's largest activity of the year, its annual show, was an all-round success; people had clearly enjoyed their visit and the profit on the show account of almost £3,500 was the largest since the Society had moved to Llanelwedd. Anxious to dispel the sense of defeatism amongst members, at the March 1973 meeting of the Board of Management, officials pointed to the importance of the Society's accounts being presented to members in the correct perspective so that positive achievements were highlighted. Spirits were to be further lifted by the successful show of summer 1973, when attendance rose by almost 10,000 over the previous year to 77,024, the third highest ever for a Royal Welsh three-day show. Surplus of income over expenditure on the show account was a handsome £27,000! Officials exuded confidence that the show was now established and accepted generally, though they heeded the advice of Charles Quant, writing in the *Liverpool Daily Post,* that there was a danger the Society might be over-optimistic following the success of the show.

His were wise words, for the Society's finances were not yet on a firm footing and a desperate temporary situation was indeed to embarrass it in 1974. That year gloom was once again spread by the overall accounts, for as a result of the substantial drop in profit on the show account – from £26,741 to £13,579 – the £6,673 profit for the year 1973 was turned into a loss on the year 1974 of £5,829. This was explained by rising expenditure at a time of inflationary costs, with a particularly savage 18 per cent increase due to wages on the main contractor's bill. With the benefit of hindsight, the Honourable Islwyn Davies claimed that it had been a mistake not to (as the Royal Agricultural Society of England had done) increase any of their charges for the 1974 show. On top of these disappointing overall year's working figures, however, came an embarrassing crisis when the Society embarked on capital development expenditure in 1974 on two items, namely, the purchase of horseboxes, costing £33,000, and of the railway station yard, costing £14,000. Although these purchases were justified in the light of current inflation, the Board had gone ahead with purchasing 300 horseboxes on the strength of promised donorship which did not come to fruition, so landing the Society in difficulties. At this critical moment, the Society was rescued from its temporary plight by the faith shown in it by Emrys Evans, regional director of the Midland Bank's headquarters at Cardiff and a co-opted member of the Society's Board of Management since 1973, who persuaded Midland Bank Ltd to 'exceptionally agree' to increase the overdraft limit substantially – to £100,000 – for a short period in the light of collateral. The Society owed much to this accommodation, and acknowledgement of lasting gratitude was made by the ever-gracious John Wigley, writing in the *Journal* in 1987. The sad episode saw some criticism being directed at the Board of Management for entering into purchases until the money was available. Bitter in his reproach of the Board's 'high handed attitudes' was Peter Manning of Devon, vice-chairman of the Livestock Committee and the inspiration behind the Horse Box Erection Programme, who angrily cut all his links with the Society because of the shoddy way that he felt he had been treated by the Board.

These years between 1963 and the mid-1970s were a time of deep-seated financial insecurity in the annals of the Society. Indeed, the first decade at Llanelwedd resulted in a net loss of over £76,000. There had been substantial criticism from north Wales from the outset at the decision to move to the permanent site, but the early financial setbacks were causing 'destructive criticism' among the Society's membership generally by autumn 1967, in the face of which Lieutenant Colonel FitzHugh sought to rally support at the annual general meeting in October. Later on, at the annual general meeting of April 1973, the finance officer was to refer to the constantly encountered problem of defeatism among members.

Part of the explanation for the want of confidence and approval among the membership at the way things were going doubtless lay in the fact that for the first ten years at Llanelwedd much of the capital expenditure on the site was largely hidden. Although solid, if somewhat slow, progress was being made to develop the showground, much of the £162,000 invested in the site by the end of 1972 had been spent on items like ditching, drainage, water mains and pipe laying, sewerage and electricity installations, which, though crucial, did not effect a spectacular change in the appearance of the ground and as such were not fully appreciated by visiting members and the general public. It was, in fact, only in 1972 and 1973, after the basic work on essential services had been completed, that the Society was able to embark on a building programme. Certainly the organizers were keen to press on with this in order to reduce very considerably the annual cost of showground erections. It was here that the feature counties came to play the vital role which Dr Alban Davies had called for in 1967: the Society encouraged them to aim at raising sufficient funds to provide a specific item, be it building or equipment, which would reduce annual erection costs on the showground. County committees

47. Early work on the new site at Llanelwedd.

were given advice so that a planned building programme was implemented, towards which goal Messrs Alex Gordon and Partners, Cardiff, were hired as consultant architects to the Society in 1973. An approach was normally made to the county council and other councils to contribute towards both the county exhibit on the showground and the permanent project to com-memorate the year when the particular county was featured.

The first significant structure was the Clwyd Exhibition Hall, contributed by Denbigh and Flint and opened at the 1972 show. Carmarthenshire set itself a target of £50,000 and provided a new Livestock Building for the 1973 show to house the cattle, a building which created a tremendous impression among the 1973 showgoers and which did much to create the new spirit said to be prevailing on that auspicious occasion. It was soon the turn of the horse exhibitors to press for permanent stables, and this led to the installation of 300 permanent boxes just in time for the 1974 show. Costing £34,000, they – for all that lack of donations towards their purchase created financial embarrassment – saved the Society at least £3,300 on the cost of staging that show. Montgomeryshire, the feature county for 1974, would pay for a new members' pavilion, which, taking nearly two years to build, was finally completed for the 1977 show. In 1975 it was the turn of South Glamorgan with funds raised for the South Glamorgan Exhibition Hall; costing £80,000 this was the most impressive building to date on the showground and one of the largest in Britain. The same year saw the appearance of a new sheep-shearing unit, a project made possible by the urgent action of eight members of the sheep-shearing committee who gave the Society an interest-free loan of £100 each. Despite the fact that the show in 1975 cost £85,000 to stage – £5,000 more than the previous year – the figure would have been astronomical had it not been for the savings on the permanent buildings recently constructed, thereby rendering it unnecessary to hire tents.

Although the 'Royal Welsh' organization has worked together throughout its history as one big extended family of officials, administrative staff, Council members, committee members, county representatives and voluntary helpers over the show period, in the years 1963–75 certain persons stood out as key contributors to the Society's standing and development. Bestriding the Society like a colossus from the beginning of the 1950s to the early 1970s was Lieutenant Colonel G. E. FitzHugh of Plas Power, Wrexham. Having already acknowledged his wise guidance during the migratory years down to 1962, it is now appropriate to highlight his continuing role at the helm of the Society after its move to Llanelwedd, first as honorary show director – a post he had held since 1952 – to 1966, then three difficult years as first chairman of the newly constituted Board of Management from January 1967, and, finally, as chairman of the Council from the close of 1969, when he took over from Brigadier Sir Michael Venables-Llewelyn, to 1972. Although never allowed to stand in the way of choir practice on a Thursday evening in Bersham Church where he was organist, his service to the Society was invaluable in terms of thought, time and energy. His approach throughout was characterized by fairness and firmness and he, above all others, shaped the Society's development in the crucial decades of the 1950s and 1960s when so much upheaval and restructuring occurred. Fittingly, he was awarded the Society's Gold Medal in 1969.

Mention was also made earlier of Sir Michael Venables-Llewelyn as chairman of the Council from 1954 until November 1969. It was generally acknowledged that he had served the Society admirably: he was a friend to everyone on the Council, was strictly impartial and always kept within the rules. On the personal level, he was a man of great kindliness.

The successor to Lieutenant Colonel FitzHugh as chairman of the Council was Colonel John Williams-Wynne, DSO, who, at the age of twenty-nine, had inherited the Peniarth estate in Merioneth in 1937. After distinguished war service in India, Ceylon and Burma, he settled at Peniarth in 1948, and as a dedicated countryman immersed himself in agriculture and the affairs of the rural community. As chairman of Council between 1972 and 1976 he was characteristically energetic, inventive and generous and made a magnificent contribution to the Society's management. Of concern to him personally was the need to reduce the number of trophies, to increase prize

48. Colonel John Williams-Wynne (third from right), chairman of the Council, officiating at the 1976 show.

money for the main classes, and to introduce novice classes to encourage beginners to try their hand at exhibiting.

Others of a lesser official status who none the less played an important part in the Society's affairs during the 1960s and early 1970s included T. H. Jones of Llandeilo, the Society's veterinary officer for twenty-one years before his retirement in 1968, and an enthusiastic recruiter of new members and the inspiration behind the county 200 Clubs; Austin Jenkins of Llandrindod Wells, chairman of the Livestock Committee, and his wife, Blanche, chairman of the Horticultural Committee, who uniquely as husband and wife were presented with the Society's Gold Medal for their outstanding services in 1972; and A. M. Jones, chairman of the Finance Committee, who, before his resignation in 1975, saved the Society thousands of pounds over the years by dealing expertly with the knotty issue of the contractors' accounts.

Recognition was made earlier of the sterling service of Arthur George. Secretary since 1948, he was to hold, later as secretary-manager, this most demanding position until his retirement in August 1973. 'Loyalty' and 'dedication' were words often used to describe him, and his guidance and forbearance amidst the upheaval of the move to Llanelwedd and the difficult years of settling down on the permanent site were of enormous benefit to the Society.

CHAPTER TWELVE

Clouds Lift Above Llanelwedd

The years from 1976 onwards were ones of massive growth and expansion, so that records were broken year upon year in terms of show attendance, livestock entries, numbers of trade stands and the like. Equally impressive was the transformation in the site itself, which became veritably 'pavilioned in splendour', the steady increase in permanent buildings and commercial stands testifying to the new-found faith in the Royal Welsh Agricultural Society as the leading institution in the affairs of rural Wales. Even the weather decided that the show deserved its support; as the caption beneath a photograph in the Society's *Journal* for 1992 quipped: 'It's not often you see umbrellas up at Llanelwedd!' Indeed, some of the acute problems that the Society would face were the result of its very success.

The turning point in the Society's fortunes came with the rising levels of attendance after the mid-1970s. Happily, the increase to 90,000 in 1975 was to be sustained, and in 1976 officials could boast of an attendance of 105,000, although they had not expected to break the 100,000-barrier for another two years or so. Various reasons were offered for such a welcome surge of visitors: at long last there was a general acceptance of the permanent site which brought with it a larger attendance from the farming community and, in particular, from north Wales; gates benefited from the greater number of holidaymakers and tourists coming to Wales; the show fell within the school holidays; the weather was favourable in 1975 and 1976; the increase in silage making also meant that the farming community completed its fodder conservation earlier; and publicity was also much improved through television advertising. We have already noted that the visible improvements to the on-site buildings from the early 1970s had generated an air of optimism and led to the feeling that the new venture was at last up and running. People, as always, liked to be associated with a winning concern. Closer inspection of attendance figures must await the next chapter, but from 1976 a fairly steady growth occurred. Significantly, the 1984 show saw an increased attendance from north-west Wales; many, indeed, attended for the first time. The magical 200,000 target was achieved in 1989 and numbers rose thereafter to average nearly 219,000 over the eleven shows from 1990 to 2000. Spirits would be lifted after the disappointment of cancellation in 2001 by the attendance of 214,798 in 2002. Even the attendance at the Royal at Stoneleigh was overtaken at Llanelwedd in 1985!

When the Society and show first came to Llanelwedd the ground was laid out for a maximum attendance of 75,000 to 80,000. After the 1976 show, thoughts naturally turned to whether the Society could cope with a possible doubling of the numbers of visitors initially envisaged. This would clearly have huge implications for the provision of catering, toilets, car parking, stock accommodation, site entrances, caravan parking and trade stands if the quality of services to members, exhibitors and the general public was to be maintained and enhanced. To date, big capital projects had been paid for from feature county funds, and sponsorship and funding from external bodies, but projects such as new entrances and roadways, widening of avenues, extending of car parking facilities, drains and so on would have to be financed from the Society's own funds. Thus it was important for the show to make a reasonable profit which could then be ploughed back into the development work necessary to maintain and improve facilities for the rapidly growing number of visitors and competitors. Consequently, a policy of reinvestment was rigorously adopted by the chairman of the Board of Management, the Honourable Islwyn Davies, during his term of office between 1972 and 1986. After the deficits at Llanelwedd between 1963 and 1974, profits over the years from 1975 onwards were a gladly welcomed experience. The year 1977 was a landmark, for it saw a turn in the tide in the overall profit-and-loss situation on the Society's show and management account transactions since the move to Llanelwedd; up until the end of 1976 the overall position of profits and losses stood at a loss of £27,193 and the 1977 profit turned this overall loss into a surplus of £42,729. The recurring sizeable annual surpluses – shown in Figure 1 in Appendix 3 – even if they were always frustratingly whittled down by remorseless rising costs of show expenditure, went into reinvestment. By making steady improvements to site facilities, problems were eventually overcome; following the building of three new toilet blocks for the 1988 show, that event was the first occasion on which the Society did not receive complaints about the toilets!

Recession set in at the close of the 1980s, and by late 1990 the farming industry faced a difficult time such as had not been experienced since the 1930s. It meant that the Board of Management in 1992, under its strong chairman, Lloyd FitzHugh, was very cautious in its consideration of new building projects despite the pressure from those who believed that they should have new or better accommodation. A certain tension understandably crept in. Members, especially those with sectional interests, were asked not to pressurize the Board, for improvements could only be made as far as resources permitted and it had to be remembered that the surplus for activities in 1991 was the lowest since 1983. Assurances were given that with continued successful shows, and in the fullness of time, the required improvements would be carried out. On the return of better surpluses in 1993 and especially 1994, the chairman was to remind the annual general meeting in 1995 that the continued upward trend in the show returns was vital if the ever-increasing demands for further improvements to the showground were to be met.

Despite the policy of reinvestment of surpluses on the show and management accounts, the site would have been far less developed and attractive had it not been for the continued raising of large sums of money towards financing favoured showground projects by the feature counties. It was imaginatively and intuitively grasped at the outset of the move to Llanelwedd – not least by Lieutenant Colonel FitzHugh – that the rotating feature counties were the solution to harnessing within the one-centre show the fierce loyalties and competitive energies of the earlier county-based shows, and so successful did the unique feature-county system become that it was looked upon with envy by other big British agricultural societies. Not only did the system provide invaluable financial help, but it acted also as a vital link between Llanelwedd and the counties and did more than anything else to secure and promote the acceptance of Llanelwedd as the home base for the Society and its show. Its importance in this last respect was demonstrated in 1981 during Clwyd's feature year when the Society recognized that the year had served a very useful purpose in breaking down barriers; the Society and its show had been effectively brought to the Clwyd people and the reverse was also the case. Tremendous efforts were put in by each feature county in its particular year in preparing for its exhibition at the show – thereby bringing something different to the show each year – and in raising money towards funding its favoured project by means of a host of activities which brought the county farming community together. Already by 1975, we have shown, the Clwyd Exhibition Hall in 1972, the Carmarthenshire Livestock Building in 1973, the Montgomeryshire 1974 project – a new members' pavilion which was not completed until 1977 – and the South Glamorgan Hall in 1975, betokened the crucial role played by the fund-raising efforts of the feature counties. Future fund-raising was to scale ever increasing heights, so that, whereas sums of around £40,000 to £60,000 were raised in the late 1970s, from the late 1980s amounts of over £100,000 were being achieved, with Carmarthen-Dyfed contributing the then record of £128,000 in 1987. Tellingly, the county, inspired by its president W. J. Hinds and justifiably priding itself on setting records, was determined to emulate the Ceredigion-Dyfed figure of £104,000 raised in 1983 under the inspirational leadership of Geraint Howells. In another phenomenal effort, the Ceredigion 1995 fund reached £210,000. So expensive were some of the projects that it meant, as we shall see, that a few feature counties agreed to pool their efforts to fund certain major developments. Certainly, the Society's officials were keenly aware that the feature county contributions were the very life blood of the Society's ongoing development. Financial contributions from the county and district councils played a vital part in boosting a feature county's fund, and their promised contributions also acted as a catalyst in stimulating the feature committee's efforts. Crucial, too, was the high-calibre membership of the advisory committees of the different feature counties under the vigorous and generous leadership of the Society's president, drawn for that year from the county featured. (For a list of the Society's presidents throughout its first hundred years, see Appendix 9.)

Much of what strikes the visitor to the showground today in the way of permanent buildings was only made possible through this sort of funding. The efforts of the feature counties were indeed vital in the face of the rising costs of capital projects from the late 1970s. As chairman of the Board of Management, the Honourable Islwyn Davies noted at the annual general meeting of 1978, 'it was essential to press on as quickly as possible with any remaining capital projects before costs got so astronomical that further development would be out of the Society's reach'. In 1976, it was Radnor's turn as feature county and the £40,000 raised went on constructing an extra span on the Carmarthen Building, which housed the cattle exhibits, and on financing a First Aid Centre. Conversion during 1976–7 of the Llanelwedd Hall to the headquarters office with a conference room and members' rest room on the first floor was the project chosen by West Glamorgan in 1977. Generously changing their own choice of project to accord with the Society's wishes, Merioneth-Gwynedd in 1978 provided a new sheep accommodation unit and, with the aid of a grant from Gwynedd County Council, £6,000-worth of bilingual signs for the showground. Following the Society's momentous decision in 1978 to build a new grandstand for the main ring – thereby avoiding the hiring of a grandstand for the three days at a current cost of £4,500 – Pembroke and Brecknock, as feature counties for 1979 and 1980 respectively, agreed to allow their funds to go towards the Grandstand Project, costing £264,000 and completed in time for the 1980 show after its official opening by the Rt. Hon. Lord Cledwyn of Penrhos on 7 July. The following year, 1981, saw Princess Anne, who was visiting the show, open the new president's pavilion, built with the appeal fund of Clwyd 1981. In embarking upon its biggest single undertaking to date, the much-needed herdsmen's residential accommodation complex, which cost £400,000, the Society was greatly aided by the pooled monies of Mid Glamorgan 1982, Caernarfon-Gwynedd 1984, Gwent 1985 and Anglesey-Gwynedd 1986. Help, too, was received from the Rank Foundation (£15,000) and the Wales Tourist

Board (£75,000), while the Society itself contributed £50,000. Such a tremendous effort culminated in Neuadd Henllan being formally opened by the prince of Wales in autumn 1985. While the fund-raising activities of Ceredigion-Dyfed 1983, totalling an impressive £104,000, made the International Pavilion a viable project, an important contribution was also made by Radnor-Powys 1988, who agreed that their target sum of £75,000 would go towards the ground floor of the building which accommodated the press and awards/ trophies. Accordingly, the overseas section of the first floor was styled 'Neuadd Ceredigion-Dyfed' and the ground floor the 'Radnor-Powys Suite'. Once again, the Society had to seek contributions from sponsors towards the costs of building the International Pavilion. The rejection of the Society's application for grant aid from the Wales Tourist Board meant that the project, which cost in the region of £300,000, was deferred. However, the pavilion was opened on 2 July 1987 by international opera star, Sir Geraint Evans, a most appropriate celebration of the Society's twenty-fifth year at Llanelwedd.

As feature county in 1987 Carmarthen-Dyfed raised, as we have noted, £128,000 which was allocated towards the Carmarthen-Dyfed Cattle Building extension, a project deferred until after the 1988 show and opened the following year. Having raised the sum of £145,000, South Glamorgan 1989 adopted a permanent Royal Welsh Food Hall as its project so that the quality foods of Wales might be shown to their best advantage at the show. The costs of the Food Hall, however, required additional, substantial sponsorship and meant delaying the implementation of the project. The upshot was that the Food Hall was built for the 1992 show by the Development Board for Rural Wales which let it to the Society until the latter could afford to purchase the building. Strong support was forthcoming from the Society for this venture, which was a major asset for both the show and the Winter Fair, a new one-day event which began in December 1990. It was also a prestigious building which did much to improve the appearance of the top end of the showground. For their part, Merioneth-Gwynedd 1990, Pembroke-Dyfed 1991 and Montgomery-Powys 1992 earmarked their appeal funds for the much-needed Stockmen's Pavilion, which cost over £300,000 and was to be called Hafod a Hendre. It was West Glamorgan's turn as feature county in 1993; its advisory committee set out to raise £170,000 to construct an educational building alongside the show ring next to the S4C pavilion. This was achieved with the help of West Glamorgan County Council and the WJEC, which would manage the centre. Clwyd 1994 decided to allocate its fund towards helping with the purchase of the Food Hall, the £120,000 or so raised contributing a sizeable portion of the total cost of £440,000. We have remarked on the phenomenal sum of £210,000 raised by Ceredigion 1995, money that was allocated towards the first phase of a new horse complex, which would take the form of an umbrella building containing eighty stables. This marvellous new facility was opened at the 1996 show by Tom Evans, the 1995 president.

At no time perhaps had fund-raising proved so valuable as in the contribution made by the feature counties between 1996 and 2000 towards the new Livestock Complex. Continuing with the rota system under the new unitary authorities from 1996, Caernarfon 1996, Brecknock 1997, Anglesey 1998, Glamorgan 1999 and Radnor 2000 all decided to raise money for the prestigious Royal Welsh Exhibition Centre, an ambitious project – but a potentially lucrative asset as an all-season facility – at a cost of £2.2 million. Taking the momentous step in early 1999 to proceed with this single largest investment ever undertaken by the Society, the Board of Management, encouraged by its dynamic chairman, Dr Emrys Evans, was strongly influenced by the need to be seen to be supporting the stricken farming community. Doubtless it was strengthened in its resolve by the loyalty and commitment of the five feature counties, who by the end of 2000 had raised the impressive sum of £666,831 towards the project. The will to persist was crucially helped, too, by the Society's securing a £500,000 European grant towards the project, announced in September 1998, and by the willingness of Hoechst Roussel Vet to sponsor to the sum of £25,000 per annum over a three-year period. The prince of Wales officially opened the Royal Welsh

49. The chairman of the Board of Management, Dr Emrys Evans, and HRH The Prince of Wales at the opening of the Royal Welsh Exhibition Centre.

Exhibition Centre during what proved to be a truly eventful visit to the show in 2000. Meanwhile, another landmark had been established in 1998 with the opening of Tŷ Ynys Môn at the show – an attractive information building with its four-sided clock and handsome weathervane that was largely financed by the Isle of Anglesey County Council, which proved to be extremely supportive of the feature county. It is only fair to conclude that the feature-county system has been the main reason for the Society's continued success since its transfer to Llanelwedd.

Financial backing, too, from the beginning of the 1980s came increasingly from sponsorship. Growing awareness was expressed by officials in 1977 that not only was greater show sponsorship desirable but that the Society should also attract sponsorship for major projects from commercial companies. With growing sponsorship of classes at the show – as was then happening in the instance of the big banks like Barclays and the National Westminster which decided to join the Midland Bank Limited in sponsoring the 1978 show – the Society would be able to earmark the money it was saving to help carry out plans for developing the showground. By summer 1980 there was general agreement that sponsorship, standing in the region of £6,000, was paltry when compared with that received by other major societies and that it behoved the Society to become far more aggressive in this field. This was deemed to be all the more pertinent given that in the present financial situation feature county committees would struggle to raise such handsome totals as those achieved over the past few years. Accordingly, a new Sponsorship Sub-Committee was set up in September 1980 under the chairmanship of A. B. Turnbull, and Lieutenant Colonel Desmond Evans was appointed as part-time secretary for the sponsorship scheme. Proper care was taken from the outset in planning the acknowledgement packages which sponsors received in return for their support. Desmond Evans's industriousness was to see it mount over the years, so that the sponsorship of nearly £23,000 attracted in 1981 would by his retirement rise to a record total of over £100,000 for the 1988 show. Sponsorship continued to climb until 1991, reaching a record of over £127,000 in that year, and, after disappointing totals in 1992 and 1993, surged forward from 1994 when it became the responsibility of the secretary, Peter Guthrie, and his team. Sponsorship attracted for the 1994 show totalled in cash and in kind £211,000 which, despite the difficult times in the agricultural industry, was to climb to a record £317,766 in 2000. There is no doubt that the Society benefited enormously from Peter Guthrie's unstinting efforts on its behalf.

Apart from the valuable show sponsorship, capital developments were also helped by sponsorships, of which just a few can be mentioned. Completion of the second ring control tower for the 1987 show benefited from sponsorship of £10,000 from the Milk Marketing Board and £7,500 from M and M

Timber Ltd; the extension of the Sheep Shearing Unit, opened in December 1988, owed much to the cash sponsorship attracted by the energetic president, Verney Pugh; a new Communications Building was used for the first time at the 1996 show, thanks to the sponsorship of Land Rover; the refurbished International Pavilion opened at the 1998 show was sponsored by Marks and Spencer; and, in response to the persuasive gifts of chairman of the Board, Dr Emrys Evans, a number of generous donations were forthcoming from various commercial firms, trusts, foundations and breed societies towards helping the Society make up the shortfall in funding the costly Royal Welsh Exhibition Centre. Dr Emrys Evans's efforts had similarly been vital in attracting sponsorship for the 1979 Grandstand Project. Special mention must be made too of the Midland Bank's sponsorship not only of the summer show but also of the Winter Fair from 1990 onwards, in addition to its making available, at difficult times, additional borrowing facilities on a temporary basis. Indeed, from its earliest days the Midland Bank (now HSBC) has been the Society's most generous financial backer as well as provider of its succession of honorary treasurers. Finally, grant aid from the 1980s from organizations such as the Development Board for Rural Wales, the Mid Wales Development Board, the Wales Tourist Board, the Welsh Office and European Funding (European Agricultural Guidance and Guarantee Fund) was of crucial importance in sustaining the ongoing capital development of the site. Of course, not all bids for funding were successful: notably, the substantial bid for £7.5 million to the Millennium Commission towards helping turn the showground into a centre of excellence for rural Wales was myopically rejected.

As well as the permanent buildings erected by the Society itself were those built by the numerous organizations, whose decision to take permanent stand space reflected their faith in the Society's future. As a statement of its confidence in the Llanelwedd project the Midland Bank erected a permanent building on the site in time for the 1978 show; it was one of the first permanent buildings to be erected by an external organization and was the first to be built since the Society introduced its guidelines for the overall development of the showground. Other organizations were quick to follow suit. That keen observer of the Society, Roland Brooks of the *Western Mail*, was struck by the number that had gone up in the twelve months preceding the 1981 show and he saw them as improving the appearance of the showground. Included in the list were two banks, Barclays and National Westminster, the NFU Mutual Insurance Society – next to the National Farmers' Union building, erected in time for the previous year's show – the Farmers' Union of Wales, the Welsh Agricultural Organisation Society, and the Powys County Agricultural Education Unit, which housed the Young Farmers' Club Wales Office, which, importantly, was used throughout the year. By the 1985 show some twenty-three organizations had taken permanent stand space at the showground. Although others would follow, permanent building development by outside organizations slowed down from the mid-1980s.

The Llanelwedd site has a wonderful physical setting and its visual qualities have been enhanced since the late 1970s by a progressive policy of constructing buildings that blend with the natural environment. No one was more conscious of the need to preserve the beauty of the site than the newly appointed chairman of Council, Lord Gibson-Watt, and this meant that the Society was strict about the architecture of permanent buildings, the improvement of the trees and keeping the showground as free from litter as possible. Such a laudable concern for the aesthetic impact of the site led to the setting up of a Long-term Planning Committee in late 1977, which drew up a list of guidelines for any potential designers when preparing a scheme for the showground. Professor John Eynon was brought in by Lord Gibson-Watt as the Society's architectural advisor, and much credit for the high quality of the permanent development from the mid-1970s lies with him. Lord Gibson-Watt's desire to keep the showground clean and tidy was also to be achieved, thanks to the outside staff, under the estates officer, Brian Waller, who moved from Hartlepool to take up his post in January 1980.

A serious threat to the showground and its development was posed in the early 1990s by the Builth Wells by-pass proposals, for both red and blue routes proposed in autumn 1991 would have had disastrous effects on vehicle parking and on the caravan parks, as well as stifling any future development of the showground. Strong representations to the Welsh Office saved the day: the adoption in 1995 of a revised 'orange' option as the official route offered the Society all the benefits of a by-pass without the major disadvantages that would have come about if one of the other options had been chosen.

Improvements to the site brought about by increased permanent buildings and amenities were, of course, in themselves a valuable means of increasing revenue in the out-of-show season. It was the possibility of greater rental revenue coupled with rising sponsorship that was perceived in 1980 as the means to offset the rising costs that had to be borne by the Society. Fuller use of the site throughout the year was clearly in evidence during 1983 when there were more than sixty events at Llanelwedd, involving some 120 days' use of different facilities. The rental revenue from these activities now approached £30,000, compared with the £6,000–£7,000 yielded by thirty-six events in 1978 and £8,000 in 1980. The holding of these events was, in turn, made possible by much improved maintenance of the estate – for after each event there was much to clear up. Predictably, with the coming into use from 1985 of the stockmen's residential block, by 1991 showground events had increased still further and the Society was receiving more revenue from this source. Given that surpluses decreased in 1991 and 1992 in the face of recession and the ever increasing costs of staging the show, officials became increasingly anxious to maximize the all-year-round use of showground facilities to produce even greater returns. By early 1993 it was felt that three or four major non-agricultural events were needed to secure the level of income necessary to protect the Society's position as returns from the show itself declined. One such event, the National Eisteddfod, was held there the following August. Notwithstanding the increased surpluses on the Society's activities from 1993, greater use of showground facilities, together with stringent controls over its financial affairs, was the golden key which enabled the Society to go from strength to strength, its honorary treasurer, Richard Moseley, observed in 1995. Great faith was to be placed in the potential of the new Exhibition Centre for attracting national events to the site. Those present at the 1999 annual general meeting were told of the showground's out-of-show events increasing, particularly in relation to conferences, seminars and meetings.

Always crucial to the Society's financial health was the size of membership, and throughout its history down to the move to Llanelwedd and for some afterwards low membership had induced a condition of anaemia. Indeed, at the outset of the move to Llanelwedd the paltry membership became veritably life-threatening, and in spite of the initial increase in numbers following the recruitment launch in 1964, membership by the early 1970s had once again slipped back and lay becalmed in the doldrums. As late as February 1977, fully-paid-up members numbered merely around 4,000. Early that year, ever aware of the usefulness of income from membership subscriptions as a sort of insurance policy, the Board of Management appointed a Membership Sub-Committee under the chairmanship of Tudor Davies. With the increase in subscription rates to £10 single and £15 family membership in February 1981 there came, too, a new and more efficient system of payment by direct debiting. Membership rose appreciably in the early 1980s, increasing from 6,824 in 1981 to 9,214 by 1983, and from this time onwards there would be periodic discussion as to the wisdom or otherwise of placing a ceiling on membership, given that it was exerting pressure on the facilities available during the show. No such imposition occurred, however, in spite of mounting membership and so there was greater pressure than ever on facilities in the late 1980s, when membership rose from 10,326 in 1987 to an all-time peak of 13,884 in 1991. The following year, 1992, numbers fell to 11,713, after which a modest recovery occurred to reach 12,454 in 1996. Despite the crisis within the industry, the membership remained commendably strong, settling down to a figure of around 12,000 down to the end of the century.

Appendix Three demonstrates (among other things) the rise in subscription income over these years. Of course, to a significant extent this reflected the rise in subscription rates. Although these were generally increased every three years in the 1990s to take account of the ever rising costs of running the Society and the show, membership rates in the 1980s and 1990s were nevertheless the lowest of the major societies in the United Kingdom. To encourage the involvement of young people in the Society's affairs, a Junior Committee had been set up in 1988 and in 1990 a junior category of membership was introduced to cater for young people up to the age of twenty-one. Taking 1985 as a typical year, membership across the Welsh counties was strongest in Cardiganshire (1,585), Breconshire (1,494), Radnorshire (1,193) and Glamorgan (1,070) while, in English counties, it is hardly surprising that the two border ones of Herefordshire (209) and Shropshire (108) boasted the strongest membership. The majority of the members were farmers and landowners, their activities representing agriculture, horticulture, forestry, rural activities and associated businesses.

Overall, the years from the mid-1970s, apart from the dreadful year 2001, were increasingly successful ones for the Society. The healthy and rising overall surpluses on the annual workings (see Appendix Three) stemmed not only from show profits but, as we have seen, from attracting additional rent through the greater use of the showground in out-of-show periods, from increases in short-term investment interest and from membership subscriptions. Surpluses, swollen by the feature county contributors, were ploughed back into site development and were crucial in transforming the showground into the beautiful site of today. The total £250,313 raised by the feature counties down to 1977 was to exceed £1 million by 1988 and to climb to a staggering £2,711,000 by 2000. All this capital expenditure, boosted by sponsorship money and grants, meant that the value of the site grew enormously. Whereas the value of the freehold property shown at cost stood at £511,711 in 1976, by 1986 it had appreciated to £2,100,000, mounting thereafter to over £4 million by 1994 and, by the close of 2000, to £6,677,984. Similarly, its net assets of £554,000 in 1976 had grown to £2,454,000 in 1986, rising thereafter to £5,530,000 in 1996 and to more than £8 million by 2000.

All this cascading success from the mid-1970s did not mean that the Society encountered no problems along the way. In the first place, its officials always had to keep a careful eye on the escalating costs of running both the Society and the show. Whereas putting on the 1976 show had cost £131,000, the cost of staging it had risen to £491,000 in 1986; the £1 million mark was overtaken in 1991 and by the year 2000 show expenditure had climbed to £1,500,957! Healthy show surpluses were always made, but in order to achieve good profits officials had no choice but to make certain revenue adjustments like increasing the charges for gate admissions, catalogues and official programmes and trade stand fees; admission charges for 1989, for example, were raised by approximately 20 per cent. The early years of the 1990s, in particular, were worrying ones, when, in the face of the depressed economic situation and the mounting costs of staging the show, the Finance Committee was forced to identify areas where savings could be made, such as catering and essential services. Another way in which the Society reduced show costs was through the appointment of an estates officer in 1980 – the inflation of the late 1970s having caused concern – whose small workforce, through producing items at less cost, eased the burden of having to meet the annual increases on the contractor's price schedule of 15 per cent. As in earlier times, there was still a need for caution and vigilance, through carefully costed budgets and monthly inspections, with prudence and consolidation constantly being urged by the honorary treasurer, Richard (Dick) Moseley.

Expansion, too, brought problems in its wake. Confronted with the immense difficulties of traffic flow, car parking, and gate and entrance organization, which had occasioned serious discomfort and inconvenience at the 1978 show, the Society took urgent remedial action from the close of the 1970s. For instance, it abolished car parking charges in 1979, widened entrances to, and exits from, car parks, secured more fields for parking areas and – in time for the 1980 show – improved the entrance roadway behind Penmaenau Farm to the top fields. For its part, the 1979 show brought home to Society officials the inadequacy of the three services – sewerage, water and electricity – and speedy action was taken to improve them in readiness for the following year's event. The problem of an inadequate water supply would again take centre stage in the late 1980s, when the Society boldly committed itself in 1988 to funding a new water project which would cost £480,000 over three years, the balance being met by Welsh Water. A grant of £100,000 by the Welsh Office and concessions on Welsh Water accounts for a twenty-year period meant that the Society's net total investment would be less than £300,000. However costly, the project, which was commissioned in 1993, was essential towards safeguarding the continued well-being of the show.

Vastly increased attendance and livestock entries were also to be a mixed blessing. The numbers visiting the show and the demands of trade stands and livestock entries by the end of the 1970s had led to uncomfortable conditions on the showground and car parks full to saturation point. Such was the limited size of the ground that visitors could enjoy the show in a relaxed way only if the attendance did not exceed about 50,000 on any one day. Faced with overcrowding, the Society had no choice but to extend the show to four days from 1981 onwards, even though influential voices in the Society's affairs opposed it and support for a four-day show at the Council meeting in November 1979 which took the decision was not overwhelming, with 48 for and 26 against. To the fore in urging the Society to take this step was Tudor Davies of Glamorgan and he received strong support from Ifor Lloyd and Alan Turnbull. When announcing this initiative in the pages of the *Journal* to members, chief executive John Wigley explained: 'We are the last of the national societies to enlarge in this way, and in doing this we hope that it will be of benefit both to competitors and the public, as there will not be so much urgency to adjudicate the sections, and the crowds will be spread out over four days.' By the mid-1980s, however, there was growing concern that even the four-day event had not truly solved the problem of overcrowding stemming from the spectacular rise in the show's popularity. As at the close of the 1970s, officials knew that one of the special appeals of the showground was that people could get round the show in a day and so, mindful of public sentiment, they were careful after 1985 not to allow the show to get bigger by pushing the perimeter fence further into the car park. It meant, however, that by the end of the decade, with more and more people and livestock, the showground was totally full. In the 1990 show limits had to be imposed on the Commercial Cattle Section and no exhibitor was allowed more than two entries per class in the Sheep Section. Even so, the 1993 show appeared to be at bursting point in the Sheep and Horses sections. As early as the 1987 show, chief executive David Walters was repeatedly asked by visitors to establish a five-day event. That possibility was mooted within the deliberations of the Society in 1994 for two reasons, namely to spread the load and to provide the Society with extra revenue. This did not happen, for discussions at county level had shown little enthusiasm for the idea and neither agricultural trade exhibitors nor livestock owners would have welcomed it. Building extra avenues at the edge of the ground, such as the one constructed for the 1994 show, was one way of easing the pressure. Alerted by Tudor Davies to their potential value, the Society's purchase of the 130-acre Wern Fawr Farm and of seven acres of glebe land in 1995 proved useful for car and caravan parking.

Heavy demands on the Society's administration, following the expansion of activity on all fronts, necessitated adjustments too. By the early 1990s it was clear that the office staff were becoming deluged with committee work, thereby denying them the time to manage the Society's affairs on a day-to-day basis. Accordingly, from January 1994 the Board of Management met monthly rather than twice a month as hitherto, and additional executive powers were given to David Walters and his staff. With this added responsibility, it was important that the senior administrative staff were allowed more time to manage the Society's affairs. This could have been achieved by reducing committee work or alternatively employing more staff. As in the past when the former had been attempted, it was evident that committees were very reluctant to reduce the number of meetings and the Board, sensitively acknowledging the importance of the voluntary nature of the Society, did not persist with its plans to prune the committee structure, instead electing to take on extra staff. These included an assistant secretary, Sheila Saer of Builth Wells, who was appointed from October 1994. At no other time would the dedication and loyalty of the staff be so severely tested than in the gloomy days of the foot and mouth crisis of 2001. The uncertainty over future prospects for continued employment produced an anxiety that had to be faced head-on by every member of the payroll. Echoing the attitude taken by the Council in late November 1939 towards shielding the staff from hardship in wartime wherever possible, the Board of Management took the view that the Llanelwedd team should be maintained for as long as possible regardless of the threat to the Society. The staff responded magnificently, giving their best throughout the long period of crisis. Fortunately, the Royal Welsh emerged from the foot and mouth outbreak in time to save the jobs of everyone and the team remained intact as one of the Society's most valuable assets.

With the exception of the foot and mouth crisis, these were all problems born of the Society's success in – at last! – commanding the support of the people of Wales and beyond. Small wonder that officials looked forward from 1999 to a relaxed easing into its centenary celebrations, with a recent outstanding president of the Society, the Honourable Shân Legge-Bourke, acting as the dynamic chairperson of the Centenary Appeal Committee. (She, together with Max Boyce and David Meredith, would launch the special appeal at the 2002 show.) Hopes of a smooth run-up to the centenary year were to be cruelly undermined, however, by the devastating outbreak of foot and mouth in 2001 which, following so closely in the wake of a prolonged period of depression in the farming industry, catastrophically harmed both financially and psychologically the farming and wider rural community of Wales. Chief executive David Walters wrote feelingly in the 2002 *Journal* of the Society: 'The year 2001 was one of the most traumatic years in the history of the Royal Welsh Agricultural Society.' No other choice lay open to the Board of Management in April but to cancel the show. Such a drastic step had been taken only once before in the midst of petrol rationing in 1948, although the Society had also withheld its show in 1938 on the occasion of the Royal Show visiting Cardiff. Not only did this cancellation in 2001 inflict a severe financial blow on the Society with a loss on the year of £390,000, but it ruinously hit the local rural economy as well. The cancellation of both the show and of seventy-four out-of-show activities meant that the local economy lost perhaps in excess of £35 million. Faced with the crippling loss which, it was foreseen, would seriously harm the Society's finances for the following three years, the chairman of the Board of Management, Dr Emrys Evans, appealed in the late summer to the minister of rural affairs of the

50. President W. J. Hinds (left) presenting the Gold Medal of the Society to the Honourable Islwyn Davies in 1987.

National Assembly for Wales, Carwyn Jones, for financial help to the tune of at least £400,000, to be paid either as a one-off contribution or in annual instalments over the following three years. News of its rejection in early 2002 naturally came as a bitter disappointment; the Society understandably felt that it was worthy of public support, given its long-standing pivotal role in fostering and promoting the Welsh farming industry and in sustaining the rural communities of mid Wales. Moreover, it had been able to assure the minister that vigorous measures had been introduced and events planned to improve its earning capacity.

It was the small staff at headquarters, the loyal membership and the unstinting service of a huge army of voluntary helpers who sat on committees, worked to raise funds in their own counties and served as stewards at the show which to a large extent ensured the Society's continuing vitality. Once again in the years from the mid-1970s, however, certain leaders played a pivotal role in its affairs. Succeeding Lieutenant Colonel FitzHugh in 1970 as chairman of the Board of Management, the Honourable Islwyn Davies, like his father Lord Davies before him, gave invaluable service to the Society during his 16-year tenure of office down to 1986. His dedication to the Society's well-being, the clear sense of the way in which it should develop, the strength of purpose, financial grasp and the genial way in which he conducted business crucially helped it to surmount the setbacks and defeatism of the early 1970s and to develop thereafter. Undeniably he inspired the Board and Council and generally got his way, which was just as well since several of the major projects on the showground, including the Grandstand, Neuadd Henllan and the International Pavilion, would not have come about without his persistence. Yet for all his forcefulness in shaping the Society's destiny in a momentous and

51. Chairman of the Board, Peter Perkins (right), with Sir Geraint Evans at the opening of the International Pavilion.

exciting period of its history, he remained modest and self-effacing; Lord Gibson-Watt cherished an abiding memory of him in 1992 selling raffle tickets in a booth by the main ring. He was fittingly presented with the Society's Gold Medal at the 1987 show and later presented with a special personal gift of a silver hip flask/fly box by the prince of Wales on behalf of the Society at the 2000 show as a mark of its appreciation of his unique service.

The three chairmen who have followed him down to the present have each in their own way given valuable service to the Society. The first, Peter Perkins, an innovative Pembrokeshire farmer, had from the mid-1970s chaired two of the Society's committees – Machinery and Trade Stands (during which time he had enormously increased income from trade stands) and Finance and General Purposes – and he chaired the Board for five years before retiring at the close of 1990. Not only was his energetic leadership grounded in wide experience of the Society's affairs, but he brought to his post also a deep knowledge of the Welsh farming industry. Similarly meticulous in the discharge of his duties as chairman of the Board between 1991 and 1998 was Lloyd FitzHugh, the son of Lieutenant Colonel G. E. FitzHugh. He, again, brought experience to this demanding position: since his first involvement in the affairs of the Society in 1972 as a steward, he went on to sit on the Board of Management and to chair in succession the Editorial and Publicity Committee and the Finance and General Purposes Committee. Suitable recognition of his eight-year leadership of the Board, a period which saw the

ABOVE 52. Chairman of the Board, Lloyd FitzHugh, with his wife Pauline, in receipt of the Society's Gold Medal at the 1998 show.

RIGHT 53. Chairman of the Board, Dr Emrys Evans, with his wife Mair (on his right), being introduced to HRH The Prince of Wales at the 1999 show.

show continue its rapid growth and ever greater use of the showground's assets during the out-of-show period, came with the bestowal on him in 1998 of the Society's Gold Medal. Both Peter Perkins, as chairman of the Board, and Lloyd FitzHugh, as chairman of the Finance and General Purposes Committee, had devoted much time in the late 1980s to reaching satisfactory negotiations with Welsh Water over the new water supply. At the helm as chairman of the Board from the end of 1998 to the present has been Dr Emrys Evans of Dinas Powys in Glamorgan, whose roots lie in the farming community of Montgomeryshire. Returning to Wales in 1972 to hold high office in the Midland Bank, he became a member of the Royal Welsh Board the following year. From then onwards, his financial and business flair and connections, as well as his prodigious energy and huge popularity, have been placed at the service of the Society, thereby enabling it to weather storms like the horsebox episode of 1974 and the South Glamorgan Hall financial crisis at the close of the same year. The Society's only way out of its difficulties in financing the South Glamorgan Hall project was to borrow £50,000 and Emrys Evans was to facilitate this loan, despite the fact that due to the overall economic situation the government of the day had placed a complete embargo on all lending from all sources at that critical time. He was also to attract sponsorship for major capital projects. During his term as chairman of the Board Dr Evans played the vital role in persuading the Society to build the

54. Lord Gibson-Watt, chairman of the Council, enjoying the humour of Lord Cledwyn at the 1986 show.

55. Meuric Rees (second from right), chairman of the Council, at the 1996 show, photographed with (from right) Lloyd FitzHugh, William Hague and Professor Eric Sunderland.

Exhibition Centre. Moreover, in the *annus miserabilis* of 2001 his shrewd judgement was to be a valuable asset.

Working closely with the chairman of the Board was the chairman of the Council. We have seen that Colonel John Williams-Wynne retired in 1976, whereupon Major Gibson-Watt of Doldowlod, Radnorshire, who was to be created a life peer in 1979, became the new chairman, a position he held for seventeen years until his retirement in 1993. His chairmanship was an outstanding one, marked by dignity and wise counsel. Much of the Society's commitment to preserving the beauty of the showground came at the prompting of this lover of the countryside and its trees. In 1993 he was awarded the Society's highest honour, the Royal Welsh Gold Medal, for his great contribution to its affairs over a period of forty-four years. Lord Gibson-Watt was succeeded by Meuric Rees of Escuan Hall, Towyn, Gwynedd, a practical agriculturalist of note whose reputation reached beyond Wales and who, in 1990, achieved the unique double honour of receiving the Society's Gold Medal as well as winning the Royal Welsh Agricultural Society Sir Bryner Jones Memorial Award. Fulfilling one of the key roles of a chairman of Council in nurturing links with the wider membership, he was always at pains to stress the Society's reliance upon the feature-county system. As vice-chairman North Wales from 1982 – following the long tenure of O. G. Thomas – Rees's counterpart for south Wales down to 1986 was Tudor Davies of Glamorgan, who had taken over in 1977 from the similarly long-serving Colonel J. J. Davis. Forthright in his opinions, Davies contributed much to the Society's affairs, among other things urging it to stamp out the abuse of free

56. John and Sally Wigley with Pugh Morgan at the opening of the International Pavilion in 1987.

57. Chief executive David Walters being introduced to Princess Alexandra at the 1997 show.

entry to the showground in the mid-1970s (he, himself, at the 1976 show confronting and turning back some of those seeking free entry) and, later in that decade, chairing the Membership Sub-Committee from 1977, and pushing for the four-day show. He and Derick Hanks had earlier been the main driving forces behind the building of the South Glamorgan Exhibition Hall. In recognition of his outstanding services, the Society presented him with its Gold Medal, the seventh to date, at the 1987 show. Followed by James Thomas as vice-chairman in 1986, Tudor Davies became an extremely effective black-bowlered chief steward and later assistant honorary director of show administration. There were also other leading figures, like Major David Fetherstonhaugh, who were involved in the running of the show itself, and their contribution will be reviewed in the next chapter.

Ensuring the smooth administration of the Society throughout the year is its chief executive. Before his retirement in 1984 John Wigley had given thirty-eight years of loyal and dedicated service to the Society, and had eventually become its secretary-manager in 1975, a position redesignated chief executive in 1977. His capacity for hard work, attention to detail, diplomacy, humour, loyalty and integrity were widely recognized by those who had worked alongside him; others, including journalists, also testified to his accessibility and courtesy. David Walters of Llangadog, like Arthur George and John Wigley before him, was a bilingual chief executive; appointed as understudy to John Wigley in January 1976 and enjoying the full support of the membership, he became chief executive in 1984. It is generally acknowledged that the

58. Richard (Dick) Moseley, the society's treasurer, welcoming the 200,000th visitor to the show in 1989.

Society has been fortunate to have him at the head of its permanent administration staff down to the close of the century and beyond. Combining youthfulness, charm, easy communication skills and an unflappable temperament with a voracious capacity for work, his star quality played an important part in promoting the Society's popularity and effectiveness. Providing valuable service, too, from his appointment in 1968 as the honorary treasurer – taking over from Gwynne Hughes – was Dick Moseley, manager of the Midland Bank, Builth Wells. His lengthy tenure of office into the new century witnessed the Society's struggle to keep afloat amidst the rapids of the late 1960s and early 1970s, and doubtless it was that searing experience which led him to prompt the Society in the later years to act cautiously.

CHAPTER THIRTEEN

The Family Show, 1963–2004

Expansion all round   Reporters had little option but to use the stock phrase 'bigger and better' when covering the farming festival at Llanelwedd year upon year from 1963. For, whatever way it was viewed, whether in terms of gross show income, attendance figures, livestock exhibits, trade stands or main ring events, the scale, variety and standards of the show simply kept on expanding and improving, so that by the 1980s it was among the most popular farming events in Britain, sometimes indeed, as in 1985 and 1986, even surpassing the popularity of the English Royal staged at Stoneleigh in Warwickshire. Table 4 strikingly reveals the phenomenal growth of the show over these years. The growth in total gross income between the 1963 show and the one in 2000 of 4,727 per cent was stupendous, though of course inflation, as in admission charges to the showground and higher entry fees, played a significant part in this. Attendance at the show in 2000 compared with that of the 1963 event had increased by 421 per cent, livestock entries by 322 per cent and the number of trade stands by 241 per cent.

After the initial depressingly low gates at Llanelwedd in the 1960s – on more than one occasion not helped by the vagaries of the harvest – numbers of visitors noticeably rose from 1973 (77,000), climbing past the 100,000 mark in 1976 (in the process breaking the previous attendance record of 102,101 set at Swansea's four-day show in 1949) and overtaking the 200,000 mark in 1989 (see Appendix Three). Certainly, as we have shown, the over-crowded nature of the ground and the subsequent pressures on facilities had much to do with the advent of the four-day show in 1981. In the days when the show was held alternately in north and south Wales it was very much a farming occasion, but after settling at Llanelwedd the show was increasingly attended by country dwellers not engaged in agriculture and by townspeople

Table 4. Categories showing the growth of the show, 1963–2000

	1963	1973	1983	1993	2000
Total gross income in £s	40,242	104,147	561,603	1,296,342	1,942,577
Total attendance	42,427	77,024	159,157	218,915	221,000
Total livestock entries	1,645	2,419	3,867	6,247	6,950
Total trade stands	293	351	687	980	1,000

as well. Here was clear indication that the event was being deliberately developed to cater for a wide range of backgrounds and interests, although, importantly, the organizers retained the policy of keeping the agricultural bias and content of the show to the fore. Throughout the Llanelwedd years it would function above all else as the shop window of Welsh agriculture and the rural economy, exhibiting and reflecting the best in Welsh agriculture and its ancillary industries.

Even so, the compelling need to boost finances saw from the early 1970s onwards the staging of ever more spectacular main ring displays and events and the bringing of a 'new look' to showground demonstrations in order to attract townspeople, especially those from the south Wales industrial area and the West Midlands. 'Something for everyone' was indeed to become the theme of all major agricultural shows in the United Kingdom from the early 1970s. Although it was a difficult balance to get right, a winning formula was achieved whereby the Royal Welsh became viewed as a good day out for the family from both rural and urban areas; if they could not always count on perfect weather, they were assured of a friendly, relaxed, informal and distinctly Welsh atmosphere within a confined showground, the like of which was not experienced at other major shows. By the late 1980s it had gained the reputation of being the finest family show in Britain. Nor must the continually improving facilities and greater comforts on offer at the showground and the television advertising of the event from the mid-1970s be overlooked when accounting for the growing numbers of expectant visitors who poured onto the showground in a manner resembling something of an annual pilgrimage.

A survey conducted at the 1990 show, involving a random sample of over 1,000 visitors, revealed that farmers and the general public supported the show in equal strength and that they regularly returned to Llanelwedd year after year. Whereas nearly half of those interviewed said that they had visited every show since 1985, farmers revealed themselves to be particular supporters, some 87 per cent of their number questioned having attended at least three since 1985. Those farmers who visited the show were engaged mainly in livestock production and they came specifically to see livestock (50 per cent) and machinery and equipment (27 per cent), whereas the general non-farming public were less specific on items they wished to see (81 per cent claimed they were 'just looking at everything'). And, of course, many came to see old friends and to renew acquaintance after a year's separation. The agricultural journalist, Claire Powell, wrote amusingly in her commentary on the 1995 show: 'There was one farmer I used to know from Knighton in Powys who said that he would walk into the showground on the first morning and get talking and by the time he'd got halfway towards the main ring the week was over!' The 1990 survey also revealed that, although visitors came from all parts of the United Kingdom and overseas, 60 per cent came from the Welsh counties of Powys (31 per cent), Dyfed (20 per cent), and Gwynedd (9 per cent). In addition, Mid Glamorgan and Clwyd contributed 6 per cent each and South and West Glamorgan and Gwent, 4 per cent each. Four-fifths of all visitors lived within a hundred miles of the showground.

Overseas visitors, comprising farmers, breeders and business people, also attended in growing numbers. At the Royal Welsh were to be seen the best horses, cattle and sheep, particularly national and border breeds, and, with the ever increasing emphasis on exports, the Society actively encouraged foreign buyers to come and see them. In particular, overseas visitors were interested in Welsh Mountain ponies and cobs, although those from New Zealand were concerned mainly with sheep. Numbers were to rise steadily from the 1960s, so that an average of 490 for each show registered at the overseas pavilion in the early 1980s. By then, the pattern over many years had remained constant, with the largest number attending from Holland, followed by Australia, New Zealand and the USA, while other visitors came from the rest of Europe, Canada, South America and Africa and just a few from the Middle and Far East. From the 1987 show the new International Pavilion would help to attract visitors. Llanelwedd in 1988 drew more than 800 overseas visitors, a considerable increase on past shows, and 900 from forty countries visited the showground in 1990, most coming from Holland, New Zealand, Germany, the USA, Australia and France. Remarkably, one lady from Holland visited her thirtieth Royal Welsh in 1995! Particularly helpful towards generating interest within overseas countries was the setting up by Simon Gittoes (of the Society's trade stand department) in the year 2000 of the Royal Welsh website.

A livestock show above all Given the constraints of climate and soil, Welsh farming has always been primarily pastoral; for instance, livestock in 1988 provided 83 per cent of all Welsh farm output. And, despite the changes forced through from the 1980s with great encouragement given to diversify to alternative crops, forestry and tourist and leisure industries, livestock production remained the key enterprise on most Welsh farms. Simply, by virtue of the prevailing conditions of climate, soil and altitude, Welsh farmers were restricted in the choice of alternative uses for their land. Throughout its history the Royal Welsh Show has reflected this basic pattern in the dominance given to livestock exhibits; as one commentator nicely expressed it, the annual event was 'the eisteddfod of the animals'. Entries of cattle, sheep, horses and, after their introduction in 1977, goats – though not pigs – were all to rise steeply over the years following 1963, so much so that the difficulty in getting round to judging them all gave the Society's officials little option but to move to a four-day show in

1981 (see Appendix Four). Continuing fast growth in livestock entries in the 1980s meant that by the closing years of that decade, with around 6,000 head of livestock of all kinds – though primarily horses – entering the showground, accommodation of the stock was posing a very real problem. The option for the Society to control entries by means of a qualifying process via other shows was closed to it because the dates of the majority of county shows fell after the Llanelwedd late-July fixture. Inevitably, the 1990 show witnessed an entry limit imposed on commercial cattle and in the Sheep Section, exhibitors being limited to two entries per class, although the first such restriction had applied to goats in 1985. Nevertheless, in the absence of any further restrictions on entries, livestock numbers mounted steadily over the course of the 1990s to reach almost 7,000 at both the 1999 and 2000 shows. The aftermath of foot and mouth would, however, see livestock entries fall to 5,539 in 2002, but happily they were to climb to 6,978 in 2003. If the Royal Welsh by the 1990s was staging one of the finest livestock shows in Europe, especially its Horses and Sheep Sections, that distinction brought problems in its wake, for saturation point had been reached with regard to accommodation.

Increased entries came partly in response to an innovation in 1971, whereby the livestock classes were replanned to allow farmers to take their stock home after thirty-six hours on the showground instead of having to stay there for the duration of the show. Single-handed farmers were thus spared the time and expense of taking too many days away from their farms. The ever-widening range of breeds and the big growth in numbers of classes within the various livestock sections likewise produced a fast growth in entries. Examples of British breeds introduced in the Cattle Section were the popular beef animal, the Lincoln Red (1977), the heavyweight South Devon (1978), the diminutive dual purpose Dexter (1982) and the Longhorn in 1992. Likewise new British sheep breeds included the Dorset Down (1968), the Hampshire Down (1971), the very old black-and-white four-horn Jacob (1972), the Bluefaced Leicester (1981), which was then being extensively used on the Welsh breed to produce the Welsh Mule, the North Country Cheviot, Shropshire and South Devon (1983), the Lleyn, the Welsh Mule and the Oxford (1984), the Exmoor Horns (1985), and the Whitefaced Woodland and the Derbyshire Gritstone in 1989, whose rams were being used on Welsh Mountain ewes to give an out-crossing of strong, sizeable vigour to the indigenous hill breeds. Welsh breeds introduced to the show were Black Welsh Mountain sheep (1964) – reintroduced after an eleven-year absence – the Welsh Hill Speckled Face sheep (1970), the Badger Faced Welsh Mountain (1978) and the Balwen Welsh Mountain (1987).

Cattle and sheep numbers were to swell, too, through the appearance in the livestock lines of new continental breeds. This came about in response to the growing number of such breeds entering the British and Welsh farming scene in the 1970s, both as pure and crossing stock, and farmers visiting the show could thereby observe their quality. Leading the 'foreigners' at the 1987 show, for instance, was the Limousin (introduced to the show as a competitive class in 1978), followed by the Charolais (introduced in 1975), which outnumbered the popular Hereford with which it was effectively competing as a crossing animal on many mid-Wales farms, and, in third place, the Simmental, first appearing at the show the previous year. Similarly with sheep, the inclusion of continental breeds in the show catalogue like the British Texel (1980), British Bleu Du Maine (1986), the Charollais (1987), the Ile de France (1988), the Rouge de l'Ouest (1989), the Berrichon du Cher (1991) and the French Salers (1992) meant that of the forty breeds at the 1996 show eight were continental. Reflecting this increase of interest in various breeds, the first autumn multi-breed show and sale of pedigree beef cattle was held at Llanelwedd in 1991.

Mention was made in Chapter Nine of the Society's concern from 1960 to modify the pattern of livestock classification in order to meet the changing requirements of the industry. A break with tradition in the Cattle Section thus came at the 1964 show with the inauguration of a class for non-pedigree cattle, this new commercial class being brought in to keep in line with modern thinking and trends. By the beginning of the 1990s cattle classes were indeed led by the commercial beef animals, the introduction of the commercial sale at the 1988 show producing a sharp rise in entries the following year. (Such sales were discontinued after the 1992 show, however, so that the fledgling Winter Fair might have the benefit of further exhibition of commercial stock shown in the summer.) We saw, too, that an earlier innovation had taken the form of carcass competitions at the 1961 show. In keeping with this continuing concern to create classes which mirrored the requirements of the trade, the Society introduced a new class in the Lamb Carcass Section in 1993, namely, for lambs not exceeding 12 kg deadweight, suitable for the Mediterranean market.

A considerable boost to the livestock sections came with the introduction of new and prestigious competitions. In recognition of the services of Lieutenant Colonel FitzHugh, the Midland Bank in 1968 sponsored the FitzHugh Perpetual Championship Trophies for the Beef and Dairy Supreme Champions. Four years later at the 1972 show a new competition, sponsored by Harlech Television (Wales) Ltd, was established for livestock exhibitors in the form of the inter-county event to decide which county came out on top in all the competitions. This bid to emphasize the importance placed on the county link and to instil enthusiasm in the counties to do better than their neighbours was an attempt to revive and extend the old pre-war

Inter-County Competition. When HTV withdrew its sponsorship, the competition was discontinued after the 1990 show, the Board of Management sensing that the competition had had its day. A new annual £40 prize came in 1977 when, in honour of his wife, Colonel John F. Williams-Wynne endowed the Margaret Williams-Wynne prize. It aimed at encouraging novices who had never been winners in the particular section of the livestock classes in which the prize was offered each year – confined, in alternate years, to Welsh Black cattle or Welsh Mountain sheep (Hill Flock section). The 1992 show saw the introduction of a prestigious Interbreed Beef Cattle Team of Five Competition, wherein each team would include at least one female. Generous National Westminster Bank sponsorship guaranteed £500 in prize money. In recognition of another valued sponsor, namely, Hoechst Animal Health, who in 1993 had agreed to be the main sponsor of the whole Sheep Section, a new supreme championship – the Champion of Champions sheep exhibit – was introduced to the 1993 show. The following year's show would see the National Westminster Bank, already the sponsor of a very large part of the Cattle Section, sponsor a new NatWest Team of Five Dairy Cattle Competition.

Under the planning and organization of the Livestock Committee, notably chaired from 1979 to 1998 by Emlyn Kinsey Pugh, each livestock section provided its own fascination and drama. In so far as cattle were concerned, whereas the first show on the permanent site saw just nine breeds attract 434 entries, the 2000 event witnessed twenty breeds on show, with an overall entry of 757. (The same number of breeds attracted only 635 entries in 2002.) The introduction of the new breeds by no means supplanted the traditional ones, which included Welsh Blacks, Shorthorns, Herefords, British Friesians, Ayrshires, Jerseys, Guernseys and Aberdeen Angus. From the outset at Llanelwedd, entries in the Hereford classes rose markedly, understandably so given that the permanent site was positioned near the home of the breed and that for many years the Hereford had been widely used in mid Wales as a crossing animal for store cattle production. Contrariwise, its distance from the Welsh Black stronghold of the north-west and the initial non-acceptance of the new permanent site at Builth meant that for the first few years farmers of that district were reluctant to compete at Llanelwedd, thereby reversing the trend of the pre-1963 years when entries of Welsh Blacks for many years had been the strongest in the cattle section. Herefords and British Friesians moved ahead at the show in the 1960s and 1970s.

However, as the initial prejudice against Llanelwedd receded and as more breeders reached the brucellosis-free stage required by the show rules, from 1972 there occurred a significant rise in entries of the native breed. Happily, the 1970s witnessed a growing interest in the Welsh Black outside Wales, a circumstance which had much to do with its good showing at Llanelwedd – not least in winning for the first time the coveted beef inter-breed championship in 1973, pushing the usual winners, Herefords, into second place. Held since 1968, the beef-breed battle in 1970 between Beef Shorthorn, Welsh Black, Aberdeen Angus and Hereford, had seen the hopeful Welsh Black breeders bitterly disappointed at the judge's decision to award victory to the Beef Shorthorns. They were to be similarly upset in 1975 when the native breed was placed second in the beef championship to the Charolais, which was making its first appearance at the Royal Welsh. Taking the main award in the Welsh Black cattle section in all three shows between 1973 and 1975 was the bull Chwaen Major 15th, bred by Huw Tudor of Towyn. After this revival in the early 1970s, Welsh Blacks continued their good showing – coming third in numbers of entries behind the Herefords and Black and White Friesians in 1981 for instance – and though continental breeds were out in strength at the 1987 show the native Welsh Blacks were numerically the strongest beef breed with seventy-one animals. Winner of the breed championship in both 1988 and 1989 – this last year seeing native Welsh Black numbers rise sharply – was D. Bennett Jenkins of Tal-y-bont, Ceredigion,

59. The Welsh Black bull, Neuadd Cawr, winner at the Royal Welsh in 1988 and 1989.

with his four-year-old bull Neuadd Cawr, which was bred by his brother Hywel, who farmed near Machynlleth. As with Welsh Black winners in earlier shows, for instance O. G. Thomas of Llannerch-y-medd, Anglesey, in 1969 and the Roberts family of Efailnewydd, Pwllheli, in 1970, breeding and showing Welsh Blacks was in the Jenkins family blood, the brothers' grandfather, J. M. Jenkins of Tal-y-bont, Ceredigion, a founder member of the Society in 1904, having been presented with a prize by the then HRH Princess Elizabeth at the Carmarthen show in 1947. Although in nine of the ten shows between 1990 and 1999 Welsh Blacks had a numerically stronger presence than the Herefords, the native stock was always outnumbered by the Holstein Friesians, often by British Limousins and sometimes by British Charolais; indeed, it was a sign of the times that at both the 1994 and 1995 events Clwyd farmer and vet, Esmor Evans, repeating his earlier win in 1989 with his Charolais cow Lappingford Tulip, took the Supreme Beef Champion title with his Charolais cow Maerdy Empress, even though stiff opposition in 1995 came from the Welsh Black bull Deiniolen Dewi, owned by John and Susan Howe of Sussex.

60. Maerdy Empress, owned by D. E. Evans, winner at the 1994 and 1995 shows.

61. The Honourable Islwyn Davies's champion Hereford bull, Sarn Eureka.

The other beef breed so popular in Wales was, of course, the Hereford, matched in the dairy section by the British Friesian, and both fielded strong displays at the show. Predictably, the Hereford breed's main awards were taken by farmers from 'over the border'. A prominent exception to this in the early years at Llanelwedd was the Welsh farming company, Cambrian Land Ltd of Berthddu, Llandinam, Montgomeryshire, owned by the Honourable Islwyn Davies. At the first show on the site he won the female championship in the Hereford cattle section – by far the largest at the show – with Sarn Curly 4th, one of a herd founded by the late Lord Davies. In 1972, as chairman of the Board, the Honourable Islwyn – fresh from winning the championship at the English Royal – took the breed's male championship with the same bull Sarn Eureka, which had the distinction of being the only male in the entire country from the leading Hereford bull, Sarn Costelloe. The strongest cattle breed in the 1987 show was the Friesian, with a total of 153 entries, and winner of the breed championship at this as in the 1986 and 1988 shows was Bryan Thomas of Tynewydd, Whitland, with his home-bred cow Lliwe Empress. Notable at the shows of the mid-1990s was the achievement of the Holstein Friesian dairy cow, Marlais Snowdrift 11th, belonging to the prestigious herd of W. J. P. Wilson and sons of Tregibby Farm, near Cardigan, in capturing the third consecutive Royal Welsh supreme dairy inter-breed championship in 1995. A similarly memorable Holstein Friesian of the mid to late 1990s was the dairy heifer/cow Glenridge Raider Cinema from the herd of R. A., J. E. Williams and Son. Imported as a yearling heifer from Canada in early 1994, she was breed champion heifer and inter-breed heifer at the 1995 show, breed champion cow and reserve inter-breed champion cow in 1996 and, most successful of all, she was judged Supreme Inter-breed Dairy Champion at the 1997 show. Winner of the Holstein Friesian section at the 1998 show was newcomer, Highwells Broker Jackie 3rd, belonging to the Jones family of Church Farm, near Magor, Monmouthshire, and she would do even better at the 1999 show by carrying off the individual dairy title, before, alas, a broken leg necessitated her being put down in the autumn. Winning was, of course, just as pleasurable for owners among the numerically less important breeds; and here strikingly successful in the late 1980s were Dyfed breeders, Len and Margaret George, who, with home-bred animals from different blood lines, won the Guernsey championship at the three shows between 1986 and 1988 and took reserve the following year.

62. The FitzHugh Championship (Dairy Breeds), sponsored by Midland Bank, at the 1999 show: Highwells Broker Jackie 3rd (left) and Glenridge Raider Cinema. In the photo are Roy Davies (South Wales regional agricultural manager, HSBC) next to Raider Cinema, and, from right to left, W. Elfed Roberts (general manager of HSBC in Wales), Fred Williams (chief executive, Semen World) and Rod Williams (North Wales regional agricultural manager, HSBC).

With so much of Welsh agriculture concentrating on sheep it is hardly surprising that – along with horses – the Sheep Section has always been to the fore at the Royal Welsh. Indeed, by the 1990s it was arguably the finest to be seen anywhere in the world, the forty-three breeds and 2,480 entries in 2000 contrasting sharply with the 443 entries spread over just thirteen breeds at the 1963 show. Wisely, hill and mountain breeds were judged on separate days from lowland ones. Besides the strong displays of Welsh Mountain and hill sheep, Border Leicester, used widely in Wales as a crossing breed with the native sheep, and also the Suffolk were numerically strong at the Llanelwedd shows. But the coming of the newer continental breeds meant that, in the 1980s, changes in entry rankings were to occur, so much so that the strongest entry in the Sheep Section in 1986 was, in fact, from one of the breeds that originated on the continent, the British Texel, which led with 118, followed by the Suffolk with 89. Its dominance at the show was further consolidated in the 1990s, when entries rose sharply to top the 300 mark on a number of occasions. Also doing well in these later years were British Bleu du Maine and Charollais. However, competing healthily with the continentals in terms of entries in the 1990s in what was an amazing comeback for the breed, was the Lleyn, which was by then proving its worth in both pedigree and commercial flocks throughout the United Kingdom

Continuing his pre-1963 run of Royal Welsh successes with the native Welsh Mountain sheep was veteran breeder John Ellis Jones of Blaen-y-cwm, Llangynog, Oswestry, who in 1963 won the ram championship in the Hill Flock competition for the fifth time, winning the Queen's Cup in the process, and in 1966 was to have his best result ever, winning about every prize in the Hill Flock section of the breed. Similarly successful in the Hill Radnor sheep competition was breeder Vivian Jones of Abergwenddwr, Erwood, Breconshire, who won the major breed award for the seventh successive time at the 1969 show. Noticeably successful, too, in the years that lay just ahead was hill farmer and breeder Sam Davies of Llangedwyn, near Oswestry, who, farming land that surrounded the Rhayader Falls, took the male championship in the Welsh Mountain sheep classes, Hill Flock section, at no fewer than six successive Royal Welsh Shows between 1971 and 1976. Perhaps surprisingly, a supreme championship for sheep was not to be introduced until 1979, the challenge trophy going to the best pair of sheep, a male and female, from any single breed. During their years of exhibiting at the Royal Welsh from the early 1980s to 1991 Ronald and Sue Jones of Menigwynion Mawr, Gors-goch, Llanybydder – an exposed upland farm – were to achieve two Supreme Championships as well as two Reserve Championships, obtaining these rare honours in addition to collecting an inter-breed championship with two reserves, winning the Shepherd's Competition (introduced in 1982)

63. Gill Wharmby (left) with champion goat Dagvill Thistle at the 2000 Royal Welsh Show.

on four occasions and, in 1986, the championship in the Lamb Hoof and Hook Competition! Twice successful in taking the Champion of Champions Sheep Award in the 1990s were W. H. Sinnett and sons, on both occasions winning with Suffolks.

Goats became an increasingly attractive section after competitive classes were instituted in 1977, and interest began with the introduction of new breeds such as the Golden Guernsey in 1985, Saanen, British Saanen, British Toggenburg, British Alpine and Anglo-Nubian in 1987, Angora in 1991 and Pygmy in 1999. Prominent supporters of the goat section since its inception were Gill and Dave Wharmby as both stewards and exhibitors. Their high points as competitors came at the 1998 and 2000 shows when they collected the champion dairy goat award with Dagvill Quosh and Dagvill Thistle respectively.

No such success story overall, however, can be claimed for pigs. The decline in the Welsh pig industry from the early 1960s was reflected in sluggish entries for many years, cancellations of the section recurring, too, in 1973, 1974 and 1975 because of swine vesicular disease, and again in 1991. Happily a revival of the pig classes at the show came from 1992 onwards so that by 1996 there was a thriving Pig Section at both the show and the Winter Fair. Much valuable support for the section came from Tom Evans who, farming with his sons at Troed-yr-aur, Brongest, Newcastle Emlyn, was one of the Royal Welsh's most prolific winners with his famous Goldfoot Welsh pigs from the

64. The president, the Honourable Mrs Shân Legge-Bourke, presenting Tom Evans with the 1997 Queen's Cup Award for the Champion Welsh Pig.

first show at Llanelwedd through to those of the late 1990s. His prowess was mirrored in his being awarded the Queen's Cup on two occasions. Other major winners were D. Esmor Owens of Penderi Farm, Llan-non, Llanelli, who in the twenty-two years before the 1990 show had not failed to take either the championship or reserve title in the Landrace classes, and Philip Fowlie of Anglesey, who won the inter-breed championship in 1994, 1997, 1998 and 1999.

Without doubt the show's biggest attractions since its earliest days have been the horse and pony displays. Once again, Llanelwedd was to see a huge growth in this section, with entries rising from 643 in 1963 to a phenomenal 3,241 in 2000. Such a marvellous revival in the horses section after the wane in interest in the horse in the 1950s occasioned by the attraction of the Ferguson tractor – *Y Ffyrgi Fach* – reflected the growing popularity of riding among children and young people in an increasingly affluent Britain, so that children's riding classes at the show by the late 1960s had come to occupy a prominent place. Welsh Mountain ponies, with their gentle, friendly personality, even temperament, high intelligence and alertness, were naturally equipped to becoming ideal children's riding ponies. After their decline in the post-war years, Welsh cobs, too, became increasingly popular from the 1960s, not least with foreign buyers, with the growing recognition of their superb quality as 'best ride and drive' animal in the world, their beauty, stamina, agility, easiness to handle and faithfulness rendering them ideal both for hard work and for pleasurable pursuits like hunting, driving, eventing and riding. Although the Llanelwedd shows had a good entry of hunters, hacks and Arabs, and the mighty shires, ponderous but handsome in their brasses and decorations, brought a guilty lump to the throat, nevertheless the event was not a strong attraction for some of the other heavier breeds. The main draw of the Royal Welsh for horses throughout its history has been the Welsh pony and cob, and it was within their classes, namely, Welsh Mountain ponies, Welsh ponies, Welsh ponies of cob type, and Welsh cobs, that the many increases in entries in the whole horse section at Llanelwedd were to be seen. Whereas there was an average entry per show of 307 in these combined classes over the three shows between 1963 and 1965, that average had increased to 1,627 over the three shows from 1999 to 2002. Looking at in-hand cob entries alone, the eighty-three entries for 1967 rose through the hundred barrier in 1977 to reach 106, a figure which had doubled by 1982 with the 211 entries, and thereafter simply soared to reach 410 in 1989 before arriving at some kind of plateau in the 1990s with an average entry for the shows over that decade of 490.

Much has been written about the electric atmosphere pervading the watching crowds – among them a fair sprinkling of overseas breeders – during the judging of the Welsh ponies and cobs at the show, something akin to the *hwyl* at the Cardiff Arms Park and now the Millennium Stadium on international day, observed Dr Wynne Davies, the knowledgeable historian of Welsh ponies and cobs and himself a successful pony breeder. The build-up to the event was in itself exciting; enthusiasts would follow form at local shows earlier in the year and on the Tuesdays and Wednesdays of the Royal Welsh Show week large numbers of expectant onlookers would pack around the main ring in readiness to enjoy the drama and spectacle about to unfold. The Blue Ribbon in this annual equine eisteddfod went to the cob stallion which won the coveted George Prince of Wales Cup staged on the Wednesday afternoon; a truly unique atmosphere surrounded this 'showing off' of the high-stepping, flamboyant Welsh stallions, with eager fans making their way to the grandstand when judging commenced at 8.00 a.m. in order not to miss a seat and, during the event itself, letting go of their emotions in cheering on their favourite to win the prestigious cup. The judging of these Welsh cob stallions in this cauldron of an arena would in itself often meet with less than civil dissent from the spectators, especially perhaps from among the competing breeders whose rivalry during the show season was intense.

Some outstanding cob breeders repeated their triumph at a number of Royal Welsh shows at Llanelwedd. The highly successful breeder Roscoe Lloyd had started his Derwen Stud in 1944 – it was to move in 1963 to Ynishir Farm, Pennant, Aberaeron – and continued his earlier success at Cardiff in

65. Derwen Princess, Royal Welsh female champion in 1982, and overall champion in 1983 and 1984.

66. Derwen Groten Goch, champion Welsh cob (section D) in 1986, 1990 and 1992, owned by Mr and Mrs Ifor Lloyd and Sons.

67. The early-1970s Welsh cob champion, Parc Rachel, owned by Sam Morgan, Pen-Parc, Lampeter.

1953 with Dewi Rosina by winning the award at Llanelwedd no fewer than seven times. Five such victories were carried off with two outstanding animals, namely, the jet-black mare Derwen Rosina – great-granddaughter of Dewi Rosina – at three successive shows between 1966 and 1968 (it was just as well he refused a 500-guinea offer for her after winning in 1966!) and Derwen Princess at the 1983 and 1984 events. The stud's fabulous run of successes was not over yet, for under Roscoe's son, Ifor, a further four championships were gained, with Derwen Groten Goch in 1986, 1990 and 1992 and Derwen Dameg in 1989! Another veteran Welsh cob breeder was Sam Morgan of Pen-Parc, Lampeter, whose dark bay mare Parc Rachel won the George Prince of Wales Cup in 1971 and 1972 and again in 1975. Her grand-dam Parc Lady had previously won the Royal Welsh championship four times in succession from 1958 to 1961. Out of the illustrious world-famous Llanarth stud would emerge the coal-black stallion Llanarth Flying Comet, who took the championship in 1974 for owner Pauline Taylor and again at the three

68. Royal Welsh Supreme In-hand Champion and winner of the Tom and Sprightly Perpetual Cup in the 1999 show, Fronarth Boneddiges.

69. Fronarth What Ho, famous Welsh Mountain pony stallion, and winner of eighteen first prizes at the Royal Welsh up to 1978.

shows from 1976 to 1978, by which time the stud had passed by gift of Miss Taylor in 1975 to the University College of Wales, Aberystwyth, though she could not bear to part with early champion Llanarth Braint, who was the grandfather of Flying Comet. Spectacularly successful as breeders of show supreme winners, too, were the Jones family, owners of the Fronarth Stud, the family having won the Prince of Wales Cup on five occasions with Brenin Dafydd in 1970, Cyttir Telynor in 1982 and 1987, Fronarth Welsh Model in 1996 and Fronarth Boneddiges in 1999. The latter's beauty and elegance captivated the crowd and brought tears to owner Gwyn Jones's eyes when judge John Thomas took off his hat and pointed in his direction. Adding lustre to the Fronarth Stud, too, were the many wins (eighteen firsts!) at Royal Welsh Shows from the mid-1950s to 1978 of the Welsh Mountain pony stallion Fronarth What-Ho. A charmer of the crowds, he won the Tom and Sprightly Cup on five occasions.

As the star attraction on the Llanelwedd equine stage, Welsh ponies and cobs were also provided with a good supporting cast from the other classes. The growing interest in riding as a hobby ensured a strong entry for children's riding ponies, numbers increasing from a yearly average of 105 over the three shows between 1963 and 1965 to an annual one of 159 over the three events from 1993 to 1995; numbers fell away sharply from 1998, however. Likewise, the popularity of pony trekking in Wales was reflected in growing entries after classes were introduced in 1965, from an annual average of sixteen over the three shows between 1965 to 1968 to entries of thirty-two at the three between 1993 and 1995, only to slip somewhat in the late 1990s. And besides the competitive horse and pony classes – to which new ones were added such as donkeys in 1971, Shetland ponies in 1972, working hunter ponies in 1974, side saddle in 1977 and Arabs in 1978 – driving and showjumping competitions heightened the attraction of the show. Harness or driving-class entries increased from an average of just twelve at the first three shows at Llanelwedd to (including Concours d'Elegance) an average of 163 over the three events from 1999, the many competitors by this time clearly offering abundant variety of spectacle to enthusiasts. Ridden classes for Welsh cobs were introduced early on at Llanelwedd under the instigation of steward and judge, Marion Thomas, herself a winner of six supreme championships with in-hand hunters and ridden cobs, and today's showgoers may recall Jonathan Emery's Ridden Cob Champion in 1998, Calerux Boneddwr, who won the Senior Stallion, Overall Stallion and Overall Ridden Cob Championship to take the Queen's Cup. Driving had always been popular with the crowds and, from 1992, unicorns, pairs and tandems were judged in the main ring instead of the cattle ring as previously. Although entries for the jumping classes were sometimes curtailed by a clash of dates with the Royal International Horseshow or the East of England Show (Peterborough), jumping competitions were rendered more attractive from 1979 by an increase in prize money facilitated by enlarged sponsorship from Everest Double Glazing and Radio Rentals Ltd, and improvement came again in 1994 when the British Show Jumping Association cooperated with the Society in an effort to give added appeal to the scheduled classes.

Trade stands   While the show has always been primarily centred on livestock exhibits, a major attraction for many visitors was the trade stands lining the avenues in increasing numbers and variety over the years. In addition to the machinery, equipment and products for farmers and the exhibits of an educational and technical nature, a veritable cornucopia of household utensils, furniture, blankets and much besides was seductively spread out along the bustling avenues to catch the eye and to titillate the fancy. These trade stands have always provided a barometer of a show's success. Whereas the early years at Llanelwedd down to the 1972 show saw no appreciable difference in numbers of entries from pre-1963 levels, a swift rise would set in from 1973, so that the 351 of that year grew to 601 by 1979 and upwards to break the 1,000-barrier at the outset of the 1990s (see Appendix 4). Indeed, by the 1990s there were waiting lists for companies and traders wanting to set up their stands and stalls. Clearly the Society's aim at the close of the 1960s to increase income from trade stands after the disappointing amounts received in previous years had been fully realized, the section providing valuable revenue over the years from the

70. Trade stands at the 1985 Royal Welsh Show.

mid-1970s so that by the close of the 1990s it was raising upwards of £550,000 at each show and contributing around a quarter of the budgeted income for the event. After pulling out of many of Britain's leading shows because of costs in the years before 1974, farm equipment manufacturers, realizing that when trading circumstances were adverse they should spend more, not less, on publicity to maintain sales, began to return to Llanelwedd – as elsewhere – in increasing numbers in the mid-1970s, and stayed. This meant that the balance between the agricultural sector and the non-agricultural one at Llanelwedd was to be maintained – in 1984 the ratio was 49 per cent agricultural, 51 per cent non-agricultural – and this was an important factor in helping to preserve the agricultural character of the show. That the large agricultural manufacturers, including foreign makers, were stepping up their support from the late 1970s can be partly explained by the fact that these traders were pleased to deal directly with farmers, and in the process get positive enquiries and results. Their favouring the Royal Welsh on these grounds meant that Llanelwedd attracted more business than Stoneleigh in the early 1990s.

Not that everything always went smoothly in relations between the Society and this key group of supporters. At a meeting of agricultural trade-stand holders on the showground during the last morning of the 1968 show, grievances were aired and threats made to pull out if they were not remedied; the most important remedies asked for were: the resiting of the stands of the main agricultural manufacturers and dealers whereby the agricultural machinery ones were put together and given preference over those of nation-alized industries like the gas and electricity boards; the replanning of catering services; and the provision of permanent toilets. Another problem raised was the date of the show, a May event being called for. Alan Turnbull – by nature no pushover! – impressed on the Society's officials that the complaints could not be ignored and, after subsequent meetings with them, improvements were carried out in time for next year's show, though the sticky problem of the date saw no concession. Nervousness by the Society's officials, too, over switching to a four-day show in 1981 was in fact because of the trade stand holders' want of enthusiasm and their indication that continuing support would depend upon the cost-effectiveness of their exhibitions at the first extended show. Traders were also to complain at the start of the 1980s about the amount of dust rising from the avenues and settling on their displays. Overall, however, they appreciated the business opportunities offered by the Royal Welsh; one firm sold more than forty tractors at the 1993 show and, indicative of its standing as a recognized place to do business, Same Lamborghini, the international tractor manufacturer, launched two of its new models at the 1994 show. Traders' support was, in turn, of course, valued by the Society. Important officials in building up such a strong Trade Stand

Section were, among others, Peter Perkins, Christopher Beynon, Andrew Jones and Peter Evans. Nor in this respect should the marked contribution made by a number of trade stand officers in their turn be overlooked.

Other shows within the show

Although less of a draw than the Livestock and Trade Stand Sections, a wide range of displays and competitions in other sections of the show contributed to its overall appeal to people of all backgrounds and interests. The flower show, with its riot of colour, delicate blooms and wafting fragrance, has been an alluring imperative for many showgoers. The new departure, too, of using a marquee at the 1981 new four-day event, sensible in itself given the problems that arose during hot weather, provided the displays with a more natural setting than that afforded by the previous shedding. Very much the waif and stray at Llanelwedd for many years as it became shifted from one spot to another, the flower show was moved to its present location alongside Entrance B in 1990, where it blended nicely with the adjacent Conservation, Country Pursuits and Forestry area. Moreover, an old bone of contention with traders and visitors alike was removed at the 1990 show with the dropping of the former admission charge to the flower tent. Alert as always to new trends and in this instance to the increasing interest in gardening and garden planning, show officials introduced into the Horticultural Section in 1996 what would turn out to be a successful, media-covering competition for landscaped gardens, similar to those featured at Chelsea Flower Show and Hampton Court.

71. Princess Anne visits the flower show in 1981.

Whereas the mid-1960s saw park departments and trade exhibitors anxious to display, by the closing years of the decade a reversal in fortunes occurred with falling numbers of exhibitors, a reduction in prize money and a threatened scaling down of the size of the tent. Although financial constraints were to force the parks departments away, later recovery set in to such an extent that by the mid-1980s trade exhibitors had returned in overwhelming numbers and the flower tent was once again and would remain a thriving part of the Royal Welsh, featuring both its Trade and Amateur Sections for roses, fuchsias, begonias, national sweet peas and flower arrangement. In particular, the Sweet Pea and Rose Championships of Wales were much looked-forward-to competitions from their introduction in the early 1970s. Adding to the interest of the flower show was the celebrity corner introduced in 1964, where visitors could see demonstrations and listen to talks on horticultural and flower topics by internationally famous personalities. The Horticultural Section at Llanelwedd was to benefit from the calibre of its successive chairpersons, Mrs B. M. Austin Jenkins, Mrs K. Parry (later Stevenson), Ian Treseder and Dr Fred Slater.

As interest in trees and woodland grew during the later decades of the century, the Forestry Section became an increasingly popular crowd-puller. Remaining on the same part of the showground throughout, it was to enjoy improved facilities with the provision of a permanent Forestry Commission Pavilion – replacing the marquee – in 1983, the extension of space affording the opportunity for a wider range of exhibits and competitions, and the creation of an Action Ring Pavilion from 1998. Certainly Llanelwedd saw an upgrading in the Forestry Section's status at the Royal Welsh. The allocation to it of a prime and attractive open woodland setting mirrored the Society's recognition of forestry's crucial role within the Welsh rural economy as post-war afforestation continued apace. As the main timber growers in Wales by the 1960s, the Forestry Commission's exhibits did much to overcome earlier prejudices harboured against its land acquisition programme. Fast develop-ments in mechanization saw demand for trade stand space surge on the part of machine manufacturers and distributors, all anxious to exhibit and demonstrate equipment, especially chainsaws, forest tractors and mechanical timber loaders. Sadly, in the very nature of things, displays of forestry products from private estates lessened as the years passed. From its commencement at Llanelwedd, the section featured a marquee for indoor exhibitors and an open area for demonstrations, the wide range of forestry interests on display covering, besides the forest machine companies, tree nurseries, timber growers, private and public forestry sectors and educational institutions.

Apart from its important functions as an educational body and as a place to do business, both geared to promoting forestry in Wales as one of the main

72. Axe-racing at the 1990 show. 73. Pole-climbing at the 2000 show.

multi-use land-based industries, the section entertained the crowds with its skilful and robust competitions. Besides the interest afforded spectators by the Young Farmers' fencing and woodland craft competitions and by the popular Stickmaking Competition – a prominent 1990s winner was Andrew Jones of Lampeter – axe-racing events from 1966 onwards and, from 1994, the pole-climbing competition, both drawing international contestants, were among the main showground attractions. Axe-racing had been first brought to the Royal Welsh in 1966 by an Australian team, and the latter paid a return visit in 1988.

A new feature of the Royal Welsh, the Country Pursuits and Rural Crafts – including sports – Section was held in an area adjacent to the Ministry of Agriculture site at the early Llanelwedd shows, but so popular did it become that in 1978 the newly named section the Country Pursuits and Sports Area – not including rural crafts which went elsewhere – was moved to a site near Entrance B. Encouraged by the Honourable Islwyn Davies, the Society was clearly responding to the growing public participation in country pursuits. With the large pool constructed by the Welsh Water Authority as its key component, the major country pursuits such as fishing and shooting were accorded prominence; free tuition was given in casting, water-based sports were demonstrated on the pool, demonstrations were given in the old art of dressing flies, and gun-dog handling demonstrations of high standard were performed. From 1994 the finals of the Famous Grouse Welsh National Pairs casting competition were held in the Country Pursuits Area. A junior casting pool was introduced to the section in 1999.

74. Instruction in casting at the 1990 show.

75. A sailing demonstration at the 1982 show.

By the mid-1980s, moreover, and again in keeping with ongoing public attitudes, aspects of nature conservancy and wildlife features were introduced into the Country Pursuits Area. The logical outcome was the creation of the Countryside Care Section in 1990, devoted to conservation, with a new environmental pool serving as its focal point. Unstinting efforts on the part of Edward Griffiths and his team allowed for the development of a new environmental area without cost to the Society. As part of the new venture, a Countryside Care competition was introduced in 1992 to encourage school-children to respect and look after the Welsh countryside.

As at the pre-1963 shows, farriery competitions were a regular feature at Llanelwedd and were to constitute a big attraction. At first housed under canvas, they acquired a permanent building on the same spot in 1981. Appropriately, the stand was placed near the avenue which the horses took on their way to the second and main rings. An added attraction came in the 1980s with the introduction of the wrought ironwork competitions and exhibition,

76. Farriery competition at the 1988 show.

although judges were sceptical about the appearance of figurines in the competition. Three prominent contributors to the well-being of the section at Llanelwedd were Albert Lewis of Brechfa, senior steward over many years and winner of the Cookes Cup for Cart and Roadster Shoeing at Carmarthen in 1947, John Price of the Forge, Talsarn, Lampeter, who was both a steward and a demonstrator of the craft on the permanent site, and William Jones, chairman of the Farriery and Ornamental Ironwork Sub-Committee in the 1980s, and associated with the competition since 1950.

Some visitors, too, found much to interest them in the Canine and the Fur and Feather Sections. The Royal Welsh Agricultural Society, hitherto incorporating a dog show under the auspices of a local canine society or kennel association, established its own section in 1965. Steady development meant that by the mid–1980s it had become one of the largest open dog shows in Wales, attracting exhibitors from across the border as well as from all parts of the Principality. Notwithstanding its being moved from one site to another on the showground, the number of classes and entries grew over the years. By 1983 there were ninety-five classes, ten more than the previous year. Actively involved in this section from 1963 onwards were the first chairman Bill Prytherch of Caernarfon, whose long stint in office stretched until the start of the 1980s, and Trefor Evans of Builth Wells, equally long-standing chairman from 1982 and into the new millenium. Similarly, by the 1980s the Fur and Feather Section had grown into one of the best in the United Kingdom, and improvement in accommodation to cope with increased entries came about in the early years of that decade. The building of the new food hall on its site, however, necessitated a move to new accommodation on the showground. Tellingly, from 1995 the Poultry Show became known as the National Poultry Show of Wales, and by the closing years of the century, too, the rabbit show acquired 4-star status. A prominent personality within the Fur and Feather Section at the show ever since 1947, as both a highly successful exhibitor of poultry and a steward, was D. Picton Jones of Llanwnnen, Lampeter.

77. Picton Jones and his 1985 champion cockerel.

Sheep-shearing events, demanding skill, speed and stamina, became an increasingly prominent and prestigious feature at the show after 1963, the excitement and sometimes electric atmosphere bearing comparison with the pony and cob competitions. Whereas there were just five competitions in the 1962 show, by the year 2000 there were fourteen. Facilities for the competitions were improved with the provision in 1975 of the six-stand permanent shearing shed, and truly commodious was the new sheep-shearing pavilion on the showground opened in 1988 by the president, Verney Pugh, a project which owed much to his dedicated support. With the stand affording up-to-date facilities, shearing was transformed from a competitive sport to a spectator one, in the process attracting greater media coverage.

Those onlookers continued to enjoy not only the prestigious competition for the champion sheep shearer of Wales introduced in 1959 and, from 1969, the champion hand shearer of Wales competition, but also the growing number of international competitions that were staged over the years. Since it was the Royal Welsh that had suggested an international sheep-shearing competition between England, Wales and Scotland, it was favoured with the staging of the first contest at Llanelwedd in 1963. Fittingly, the main team award would be the W. J. Constable Trophy, which the Society purchased to honour the name of the late chairman of the committee which first scheduled the competition. Soon to be joined by Northern Ireland and later by the Irish Republic, the International Competition – rechristened the Five Nations Team Championship in 1987 – also provided an opportunity for finding a Champion Shearer of Great Britain. Their realization that they were fighting for a place in the Welsh team the following year to compete in the International Competition placed even greater pressure on the competing Welsh shearers at Llanelwedd. Verney Pugh's enthusiasm for shearing was again instrumental in securing the introduction at the 1990 show of a Supershear European Lamb Championship, open to the top six shearers from European, Scandinavian and Eastern Block countries, with the top prize being a return air ticket to Perth, Western Australia. A crowning achievement came in 1994 with the staging for the first time of the World Sheep Shearing Competitions, which had begun at the Royal Bath and West of England Bicentenary Show in 1977. On this prestigious occasion spectators at Llanelwedd witnessed the dramatic final between the best six shearers – Nicky Beynon (Wales), Steven Lloyd (England), Tom Wilson (Scotland), Peter Ravndal (Norway) and New Zealanders Alan McDonald and David Fagan – Beynon coming third behind McDonald in first place and twice world champion Fagan as runner up. Big prizes for the open championship attracted to Llanelwedd some of the world's top shearers, as those for the 1999 show when airline tickets to shearing competitions in South Africa and Australia and cash prizes totalling nearly £5,000 were on offer.

Of course, Welsh shearers attracted great attention in all the competitions. The 1960s witnessed the continuing battle between the Radnorshire Lloyd brothers, Isa and Sam, for the Welsh title. Sam built on his 1962 success by winning it a further five times down to the 1969 show! Notable Welsh shearers at Llanelwedd in the 1970s were Jeffrey Evans of Rhaeadr and Libanus farmer Geoff Phillips. Evans, in competing for the open championship in 1973, matched the formidable skills of New Zealander Dwight Hall. Along with the aforementioned Nicky Beynon from Gower, another champion Welsh shearer in the 1980s and 1990s was John T. L. Davies of Sennybridge, the two of them along with many others nurtured by Bryan Williams of the Society's Shearing Committee and Wales team manager for the 1995 test match with New Zealand.

We have seen earlier that, popular with the public as sheepdog trials undoubtedly were, lack of space on the show sites from the mid-1950s had

78. Nicky Beynon of Gower, winning the champion shearer of Wales competition at the 1997 show.

led to a number of cancellations of the event. Only from 1967 did they again become a regular feature of the Royal Welsh, although an alternative site for the trials was proposed and the possibility of divorcing them from the show and holding them later in the year was suggested at the Show Administration Committee in 1972. Moreover, despite the success of the Open Competition since its inception in the mid-1970s, there occurred (true to form almost!) a continuous rift between the North and South Wales Associations; whereas the former wanted a truly national trial at the annual show, the south Wales members were not in favour of running a national trial at the expense of the open one, for this would exclude competitors from outside Wales and, moreover, the open trial had only been instigated because the north Wales team had failed to turn up for the 1975 show. Facilitated by the diplomacy of Emlyn Kinsey Pugh the differences between north and south Walian competitors were overcome with the agreement to run two trials at the 1979 show, an All Wales Trial on the first day and the Open, organized on the same lines as hitherto, on the second. Happily the trials henceforth flourished, the Society introducing – the first of their kind – the Champion of Champions trials at the 1981 show, which consisted of the singles national champion from each of the four countries of the British Isles competing against each other in the main ring. Other large shows in Great Britain would follow suit.

79. Princess Margaret's visit to the Young Farmers' enclosure at the 1966 Royal Welsh Show.

We have seen that from 1949, under the aegis of Arthur George, the Young Farmers' movement in Wales attained a heightened status at the show, its newly created enclosure staging for the first time a comprehensive all-Welsh programme which would grow in variety and verve up until 1962. The high level of autonomy it enjoyed in planning its own programme meant that it was indeed a 'show within a show'. With the move to Llanelwedd there was a need to replan the organization of the movement's extensive projects at the show, with the result that responsibility was transferred from the county federation in the shire where the show was held to groups of counties in turn – in 1964, for example, Anglesey, Caernarfonshire and Merioneth combined. Each group operated on the direction and advice of the secretary of the Young Farmers' Clubs movement in Wales, who for the remainder of the 1960s would be the indefatigable Jane Davies. Later, Martin Pugh would give sterling service in promoting the Young Farmers' activities at the show. The prestige of the Young Farmers' Clubs was given an early boost when Princess Margaret visited their stand on her first visit to the showground in 1966. Over the years, the movement introduced new competitions and events, such as the Young Farmers of the Year competition in 1965, the International Young Farmers Clubs Stockjudging in 1972, the Lamb Classification Competition in 1979, the Promotion of Welsh Lamb competition in 1984, the Promotion and Marketing of a Welsh Product competition in 1986 – reflecting the needs of the hour – and the Farm Band competition in 1988, with instruments comprising farmyard articles and materials. If the movement's main purpose at the show was educational, its programme also covered a wide range and variety of entertainment and sporting activities, with the simple desire for fun animating many events. Particularly popular competitions at various times were the fashion parade, folk dancing, disco dancing, tug of war, the farm bikes event and seven-a-side rugby. By the mid-1980s many hundreds of members, after a long winter's preparation, were competing in some twenty or so activities, all keen to achieve for their county the coveted NFU Challenge Trophy. For Geraint Jones, writing in *Y Cymro* of his visit to the 1981 show, the most striking feature over the four days was the variety of activities put on by the YFCs of Wales: 'Nearly forty activities were organized by them over the four days, including many things that were nothing to do with farming, like the disco-dancing competition, the pantomime character and the fashion parade.'

Despite the close relationship between the Society and the Young Farmers' Club movement, there was some tension in the early and mid-1970s. Perhaps coming to expect too much of the 'special relationship', the Young Farmers' movement felt aggrieved over a reduction in its stand space due to the encroachment of the new South Glamorgan building, at the withdrawal of the £150-grant as well as the curtailment of provision of meals to the section because of the Society's financial difficulties, and, finally, over their exclusion from the Finance and General Purposes Committee and the Board of Management. In so far as this last upset was concerned, whereas the YFC understood itself to be a 'branch extension' of the Royal Welsh Agricultural Society and not simply another section, the Society viewed the YFC as no more than 'a most useful and essential section of the Show'. Furthermore, it felt that the movement, with 9,000 members throughout Wales, could, perhaps, do more for themselves. Some concessions were made by the Society in June 1975 when it decided that it would meet with the YFC annually in order to discuss any matters of concern and that the payment of £150 would be reinstated as long as the Young Farmers dismantled and cleared the sheep hurdles. However, the unique relationship between the Society and the Young Farmers would again be tested in the 1980s and the start of the 1990s when the unacceptable level of drinking and related unruly behaviour among the young, especially during their late evening social events, led the Society to call upon the movement to police its members in a more effective way. Nevertheless, it must be emphasized that throughout the years the Young Farmers have added much appeal and colour to show week and that this has been well recognized and valued by the Society's officials.

Furthermore, the Young Farmers played a prominent role in helping organize the Miss Royal Welsh competition which began at the 1970 show. The idea was floated in autumn 1968 by Viscountess Chetwynd of Tyn-y-coed, Arthog, Merioneth, a well-known breeder and exhibitor of Welsh ponies, who claimed that the selection of Miss Royal Welsh at the show each year would stimulate interest from every area in Wales in the winter months, and especially among the younger element, through county selection events culminating at the show. At this time of apathy towards the Society among the general public, it was anticipated that the competition would generate publicity and promote membership. From the very outset the competition was a great success, although once again there was some conflict in the first few years between the Society and the Young Farmers over the precise nature of the latter's responsibility in running the event. Much of the competition's success lay in the devoted chairmanship of the committee between 1974 and 1988 of Mrs N. S. K. Pugh. Recalled by one male judge were her words of caution that it was not livestock they were adjudicating and so they should leave the rear-quarters alone! After the judging in the afternoon, that same night the winner would attend the ball held in her honour, and later in the

80. The lucky thirteen competitors in the finals of Miss Royal Welsh, 1972.

week would add sparkle to the occasion by visiting stands and presenting prizes. Yet more colour was lent the competition from the late 1970s by the transport afforded by the vintage cars and the well-turned-out ponies and carriage. By the mid-1990s it was felt that the Miss Royal Welsh title should be changed since it degraded the competition, and the new one of Royal Welsh Lady Ambassador was substituted from the 1997 show onwards. Nicola Davies from Ceredigion was thus the last Miss Royal Welsh chosen in 1996 and her successor as the first Royal Welsh Lady Ambassador was Anwen Orrells from Montgomeryshire. From 1991 a new competition would be held at the show to select Mr YFC.

Another popular feature of the show was the Produce Section. This section had been organizing competitions since 1954 and the first working demonstration was staged at the 1961 show in Llandeilo. From 1963, the Agricultural Education exhibit became the responsibility of three different counties every year and the connection in 1961 and 1962 between the demonstrations and a specific county had to be dropped. At Llanelwedd, too, in 1963 the Produce Section was grouped with the Women's Institute exhibit and in 1965 the demonstrations were brought into the main Produce Tent, with the Bee Section taking up the previous demonstration site. This meant that the public could conveniently view the related exhibits within the Produce Section, WI exhibition and the demonstrations in the same enclosure. Moreover, in the early 1970s the Produce Section's title was changed to Produce and Handicrafts in order to take account of the wider interest in domestic crafts. Better facilities were provided from the 1979 show onwards when the Produce, Handicrafts, Honey, Women's Institute and Merched y Wawr sections were housed in the South Glamorgan Exhibition Hall.

By the mid-1980s there was an urgent realization among farmers' leaders in Wales that effective promotion and marketing was essential if the Welsh farmer was to survive, and the first signs of the Society's response to this came through the introduction of a Food from Wales exhibit in the South Glamorgan Exhibition Hall at the 1985 show, an exhibit which benefited from the organizing flair of James Thomas. It provided a suitable complement to the impressive Wales Craft Council display in the same building which was now becoming a feature of successive shows. A new dimension of the show was thereby opening up, with the two old vital staples of livestock and machinery being joined by food. Indeed, only by including in its exhibits the end product could the show be a representative shop window for Welsh farming in the changing market circumstances of the last decades of the century. The way ahead was shown by the Welsh Development Agency Welsh Food Initiative programme launched in London in June 1986. At the Royal Welsh Show of 1987 the food stands were moved from the South

Glamorgan Hall to an adjacent marquee called the Welsh Produce Centre, which contained the food exhibitions normally staged by the Society and other bodies like the WDA, Dyfed Food Initiative, Welsh Agricultural Organisation Society and Wales Craft Council. For the first time these various efforts were properly coordinated and held under the same roof, so constituting, in one commentator's headline, 'a gourmet's delight'. At subsequent shows some of the success stories among the invigorated Welsh food-producing industry would display and retail their products in the food hall. The Welsh Food Initiative was clearly bearing fruit and by the close of the decade the food hall had become one of the most popular exhibitions at the show. At the same time there was growing anticipation of the permanent food hall, a scheme made possible by the 1989 feature county's (South Glamorgan) appeal fund. That purpose-built food hall, accommodating some twenty-five exhibitors who displayed and sold a mouth-watering range of food and drink, would be ready for the 1992 show, the project having been vitally assisted by the support of the Development Board for Rural Wales, which, aiming to combine the concerns of Taste of Wales and the Welsh Food Initiative, had in 1990 sponsored the new Welsh Foods Promotions Ltd. This prestigious new building helped to promote and market the quality foods of Wales in the face of growing overseas competition, and closely involved in ensuring its success were Bill Ratcliffe, as assistant honorary director (food hall), Welsh Food Promotions Ltd (from 1994), and Welsh media personality and Taste of Wales manager, Gilli Davies.

Another way in which the show branched out from its traditional concentration on livestock and machinery was its inclusion from the late 1960s – in line with other major agricultural shows – of ever more displays and demonstrations to attract both farmers and town dwellers. Only by drawing in the latter would the Society's dire financial position be rescued, but they needed more than just farming features to attract them. The crucial change in direction came in the early 1970s after the Board of Management voted to allocate bigger sums to stage attractive displays on the second and third days. With spectacular displays like those of the Royal Artillery motorcycle display team now thrilling the spectators, agricultural writer David Lloyd reported of the 1974 show: 'While it might not please the professional showground purist it was the main ring "circus" spectacular that was the high spot for the majority to whom I spoke.' Amidst the financial embarrassment faced by the Society in 1975, the temptation to save money by not engaging a major display at that year's show was resisted by the Honourable Islwyn Davies and others who saw clearly that if the public were to be attracted, something spectacular would have to be offered. It was clear in the late 1970s, by which time perhaps as many as four-fifths of those visiting Llanelwedd consisted of the general, non-farming public, that with many perceiving the show as 'a vast live circus' its entertainment value was of major importance. The miracle was that the show organizers in the years ahead succeeded in preserving the agricultural flavour of the event as well as making it entertaining and interesting in the widest sense.

As displays grew increasingly spectacular, it became more and more difficult to stage satisfactory programmes. On occasions, clashes with other major events such as those of the Royal Tournament, the Royal International Horseshow and the East of England Show meant that certain large displays were unavailable. The desirability of offering something different each year was an obvious headache, and this was compounded by the fact that, since the Royal Welsh was the last of the major shows, some of the popular displays would have already been seen elsewhere. Moreover, by the end of the 1980s it had become evident that the Society would have to depend less and less on the military displays as most by this time had been disbanded while others were unavailable because of clashes with the Royal Tournament. Finally, there was the pressure to provide a programme that matched the public's growing expectations; thus, aware of the criticism that the previous year's programme had not been as exciting as usual, the organizers of the 1996 displays ensured that two star attractions made a welcome return, namely, the Musical Ride of the Household Cavalry and the JCB Dancing Diggers.

Such spectaculars and others – taken at random – like the Cossacks team of horsemen in 1991, French Horseball in 1988, and the Royal Danish Hussars

81. French horseball at the Royal Welsh Show, 1988.

in 1973, were seen as essential to draw the crowds, over and beyond the regular and popular features such as tug-of-war, *Y Cymro Cryfa* (Wales's Strongest Man), trotting competitions, mounted games, parades of foxhounds, motorcycle competitions and displays of veteran and vintage cars. An interesting new development came in 1985, largely owing to Alan Turnbull's imagination, with the Triumph of Henry Tudor Display to mark the 500th anniversary of the Battle of Bosworth Field, a genre that was to be repeated in 1987 with the Pageant of the Rebecca Riots of the 1840s, appropriately staged in Carmarthenshire's feature year. The Society was most fortunate to have Alan Turnbull as its programme director/chairman of the Programme Committee from 1970 to 1990. By his masterminding the show programme, arguably he set standards that were not matched by shows elsewhere in the United Kingdom. Other notable contributors to the show programme were Dillwyn Thomas (assistant honorary director, horses) and Peter Cooper.

From the late 1960s, demonstrations were also used as a means of staging a show that would be of interest to both farmers and townspeople. Whereas farmers could acquire information on new production techniques and technological developments from the demonstrations and exhibits staged at the Ministry of Agriculture site, the Milk Marketing Board stand, the breed societies' stands and the like, demonstrations of more general interest included bee-keeping, log-sawing, rural crafts, lamb cuts and their preparation, floral art, wool fashion and fly-casting. From 1970, under the encouragement of the Honourable Islwyn Davies, the Society attempted to bring a more coordinated approach to the various competitions, demonstrations and exhibitions put on in each section by adopting a particular theme for each show, and as part of this special displays were staged in the Society's Demonstration Area. In presenting these annual projects – for example, Water (1971), Farming and Recreation in Wales (1974) and Hill Farming (1978) – the approach, in keeping with the Society's independent and impartial stance, was objective. However, the initiative did not survive into the 1980s.

Growing problems within the agricultural industry from the 1970s saw Welsh farming enter a long recession in the last quarter of the century, the like of which had not been experienced since the 1930s. Amidst this gloom the annual show became increasingly a place to talk shop, a vital venue for official and semi-official discussions on the issues of the day, involving representatives from the world of politics, banking, conservation and environment, insurance, agri-economics, and farmers' unions. At the various shows of the 1980s and 1990s prominent talking points included the imposition from the mid-1980s of milk quotas, which harmed Welsh dairy farmers so deeply, promotion and marketing, the MacSharry Common Agricultural Policy reform proposals of 1991 and the wider issue of Britain's membership of the European Union, including the argument over a single currency. Thereby the show from the 1980s was serving not only as a shop window for the best livestock and produce but also as a place to debate the future of the industry within Wales.

Throughout the Society's history a special buzz paraded the showground when members of the royal family were present. In 1966 the Llanelwedd show received its first royal visit in the person of Princess Margaret and thereafter the show was honoured with a further fifteen down to and including 2002. Great excitement filled the showground on the occasion of the visit of the Queen and the Duke of Edinburgh to the 1983 show from the moment they arrived at the Gibby Gates in the Balmoral Sociable and Landau carriages, drawn by horses from Buckingham Palace stables and accompanied by mounted outriders. Although the proceedings at the end of the lunch attended by the royal couple and 375 Council members and official guests were somewhat disturbed by a US Air Force Phantom aircraft flying low over the Pavilion, the day-long visit was a memorable success, culminating in the couple being driven into the main ring in the royal carriage accompanied by loud cheering before entering the royal box to view the parade of prize-winning stock and a special parade of prize-winning Welsh pony and cob stallions. The warm welcome accorded to the Queen and the Duke was shown once again to the prince of Wales on his five visits to the show since his investiture. Indeed, with the farming industry in such disarray in the late 1990s, Prince Charles's visits to the showground in 1999, 2000 and 2001 were

82. Her Majesty the Queen and the Duke of Edinburgh at the 1983 show.

83. HRH The Prince of Wales giving the opening address at the 2001 Winter Fair, with Stanley Thomas, president.

84. HRH The Princess Royal opening the 2002 show.

morale boosters to Welsh farmers and their stricken industry, especially his address in opening the 2000 show, which was, for the first time, held in the main ring. There was a gratifying empathetic identification with the Welsh farming community in his rousing finale: 'Wales has so much in its favour, some of the finest grassland in Europe, incomparable scenery and skilled farmers. I pray with all my heart that the future may be brighter and, above all, that the farming community and all that it stands for, is supported, protected and cherished.' A farming community recovering from foot and mouth disease was heartened, too, by the magnificent support and concern shown by the Princess Royal on the occasion of her opening the 2002 show.

An important element of that farming community also cherished its native tongue and it was an achievement on the Society's part that it sought to strike a fair balance between its Welsh-speaking and English-speaking members and visitors to the show. During the pre-1963 years there had been some progress in the use of Welsh in the conduct of the Society's affairs, including bilingual ringside commentaries at the show. The 1970s, in particular, saw pressure to enhance further Welsh-language provision at the show and the Society responded positively. Sympathetic consideration was given in the mid-1970s to the introduction of Welsh at the show's opening ceremony. It came in response to farmers' complaints, *Y Cymro* for 29 July 1975 reporting thus: 'This year again, like last year, the Farmers' Union of Wales complained about the

85. The duke of Gloucester chats to the long-service medal winners at the 1982 show.

lack of use of Welsh in the opening ceremony of the Royal Welsh Agricultural Show at Builth.' The 1978 show saw the inclusion of Welsh on prize cards and, likewise, Merioneth, as feature county in 1978, stipulated that some of the money which had been collected should go towards the provision of more bilingual signs on the showground. Further progress came at the close of the 1980s when recipients of long-service awards, whose presence, as in pre-1963 days, was a special feature of each show, were given the choice of having their medals inscribed in either English or Welsh. From the 1999 show onwards, provision was made for Welsh entertainment at the bandstand each evening.

Problems nevertheless

Despite records for attendance figures and numbers of entries being broken year upon year, show organizers faced problems, not least those related to traffic, illegal entry and theft. In 1990, for instance, as many as 92 per cent reached the showground by car and ensuring ease of access to and departure from the showground posed a constant headache. Gatecrashers in the late 1960s, mainly livestock exhibitors with a vanload of people without passes, trade exhibitors and members' friends, necessitated a strict clampdown by the stewards on the gates. There was no easy solution to the problem, however, and reports of illegal entry via gates and at the machinery and livestock entrances were heard for many years to come. Above all others, it was Tudor Davies who vigorously set about putting a stop to the abuse in the mid-1970s, for in his mind show profits could only be boosted if all visitors paid for entry to the showground. A husband-and-wife security team from Bristol mounted on horses was hired at the 1977 show in an attempt to stop illegal traders, unauthorized car parking and people trying to gain entry without paying. Despite these efforts to solve the problem, organizers were to lament of the 1981 event that illegal entry had reached 'unacceptable proportions', although happily the problem diminished in the years that followed. Another matter of concern during the early shows at Llanelwedd was the high level of internal traffic on the showground, which was regarded as so serious in 1972 that Major Basil Heaton was entrusted to take steps to reduce it. Although his efforts went some considerable way towards freeing the avenues from this nuisance, the problem was still being complained about a decade later. Showground security, too, posed a problem in the 1980s as incidents of theft increased, but tighter measures meant that the situation improved from the 1987 show onwards.

Much more worrying and ominous, however, was the growing abuse of alcohol at the shows from the start of the 1980s, mainly among the younger age groups. Long and heavy drinking sessions at the various bars and the unpleasantness that can accompany them threatened to disfigure the whole event. Despite efforts to stamp out this small but unacceptable and unsettling tendency – a campaign coordinated by Robin Price of Rhiwlas, Merioneth – little success was achieved, and in September 1986 the show organizers were to observe darkly that there was 'an increasing amount of drunkenness, hooliganism, and pilfering'. Mobs of young people frequenting the stockmen's bar created much of the problem. Excessive drinking disrupted the peace and quiet of the external caravan and tent parks, and such was the deterioration in the standard of behaviour in the members' pavilion – described as a 'Wild West Saloon' – that by the close of the 1980s family members were no longer using the facility. The last circumstance would change dramatically from the 1990 show with the introduction of stricter regulations and the siting of a members' bar elsewhere. A welcome improvement came, too, from the start of the 1990s with the formation of a young people's tented village, christened at the outset Happy Valley, to the north of the showground along the main Builth Wells to Rhaeadr road. Increasingly, the YFC was given responsibility for stewarding the village. These measures served merely to nip the branches, not strike at the roots, however, and the amount of drinking and resultant boorish behaviour by young people at the show, which affected showgoers and traders alike, remained a severe problem throughout the 1990s. As if this was not enough, an added problem for show organizers was the annual pilgrimage of travellers from the beginning of the 1990s. Finding a suitable site for these travellers posed an insoluble problem in 1999 and 2000 and left the Society resigned to no other recourse than to 'batten down the hatches' and hope for the best.

Another recurring problem facing the Society was the very date of the show. At the close of the 1960s there had been a lively discussion about the relative advantage of sticking to the third week of July or moving to the third week of May, which many thought had the advantages of not interfering with vital agricultural work, affording machinery exhibitors better business and of promising good weather. Although a large majority of Board members preferred the May fixture and a referendum to members likewise saw a majority in favour of the earlier date, no decision to change was taken at this time. A decade or so later when the debate was reopened, the question was whether to hold the show in the last week of July or to stay put. Once again – in 1983 – the Society, mindful that the Welsh county shows and the National Eisteddfod were unhappy at the prospect of its switching to the final week of July, decided to continue with the third week.

Despite all-the-year-round planning and preparation, there would never be the perfect show free of unforeseen, last-minute hitches. Just a few examples have been chosen to reveal how arrangements could come unstuck from time to time. At the 1968 show a judge, Austin Jenkins, walked out of the cattle ring after an argument with the show's director, Alan Turnbull, over the timing of the judging of the FitzHugh championship trophies for the champion beef and dairy breeds. While the judge wanted the competing animals to be finally judged in the main ring the following day, Turnbull insisted that they should be judged immediately following the individual breed events. In the judge's absence from the ring the dairy breed champions continued to parade and, had he not soon returned, the show director would have secured another judge to carry on with the event! Less dramatically, the judge of the 1983 side-saddle class was unable to proceed because, expected to ride some of the horses forward, she had not brought her riding clothes. The following year witnessed show-ring drama as a Royal Marines corporal fell thirty feet into the main ring when his parachute collapsed, sustaining a broken arm and leg in the process.

Show administration

Upsets like these apart, the general standard of the show's administration was high, mirrored above all in the main livestock parade, a great show and crowd-puller in itself and enhanced over the years by an increasingly effective bilingual commentary undertaken by men like Llywelyn Phillips in the years before 1980 and, from 1972 to the present, Phillips's protégé Charles Arch of Machynlleth. Much of the show's success depended on the stewards, who

were answerable to the assistant directors of their sections. Many in the main and collecting rings had served for twenty years or more, some of them following in the footsteps of their fathers or even grandfathers. Examples of long-serving stewards were Marion Thomas, a steward in the Horses and Ponies Section from 1947 to 2000, who was the first woman to steward in the main ring and who dutifully wore a bowler hat; her husband, Harry, the proud wearer of a steward's badge and bowler hat since the 1947 show and who graduated to chief steward, parades/displays in the 1980s; Graham Rees, who similarly rendered fifty years of stewarding down to 1997, becoming in turn senior steward for Welsh cobs and chief steward of the main ring; and Ken Williams who served as a steward in the sheep-shearing section from the first show on the permanent site down to the end of the century. Certain figures were to stamp themselves on the Llanelwedd show scene year after year, crucially contributing to the smooth running of the event. Among them men like Llywelyn Phillips who, as assistant honorary director of the Cattle, Sheep and Pigs Sections in the 1970s, radiated a quiet confidence among officials that all would go smoothly, and, again, Tudor Davies who, as chief steward, show administration, did much in the 1980s to resolve difficulties with his intuitive feel for what made the show 'tick'.

Supported by his stewards and assistant honorary directors – in 1982 D. Vincent Evans of Brongest, Ceredigion, became assistant honorary director in no fewer than five sections – the honorary show director was ultimately responsible for the organization and conduct of the show. In 1967 Alan

88. Major David Fetherstonhaugh (show director from 1969 to 1989) with his wife, being introduced to Princess Alice at the 1985 show.

Turnbull was elected to this position, a post he held until becoming president in 1969, after which he became programme director up to 1990. Fittingly, a clock was placed in the main ring control building as a memorial to this legendary character, a man renowned for his great sense of humour, his genius at deflating egos, his intransigence, his delicious want of political correctness, his unwillingness to suffer fools, his gift for seeing the crux of an argument immediately, and his loyalty as a friend. Among those friends was John Thomas of the Vale of Glamorgan, whose gift for human relationships so impressed Turnbull during his presidency in 1969 that he was instrumental in making him VIP steward of the show in 1970, a position he held for decades. Alan Turnbull was succeeded as show director in 1969 by Major David Fetherstonhaugh of Plas Kinmel, Abergele, whose long tenure of office lasted until 1989. Besides his other involvements as a practical farmer, a keen devotee of country sports and horse racing and family man, he closely identified himself with the Royal Welsh Agricultural Society from his first being recruited by Lieutenant Colonel FitzHugh as a gate steward at the Caernarfon show in 1952, holding high office in the first place as assistant honorary director (administration). As show director spanning twenty-one annual events, his dedication, flair and unflappability were valuable assets in helping to meet the challenge of a fast developing showground and growing numbers of show visitors. He was awarded the Society's Gold Medal in 1989 in recognition of his outstanding services. Succeeding him was one of Britain's leading sheep

89. Verney Pugh, RWAS show director, 1989–94, and first director of the Winter Fair.

90. Harry Fetherstonhaugh, show director, escorting the Honourable Islwyn Davies from his presentation by the prince of Wales at the 2000 show of a silver hip flask/fly box for his unique contribution to the RWAS.

farmers, Verney Pugh of Cwm Whitton, Knighton, who had been assistant honorary director of the Sheep Section of the Royal Welsh for a number of years. After five years of distinguished service he was replaced in 1994 by Harry Fetherstonhaugh, the son of Major David Fetherstonhaugh. Already experienced in the workings of the Society, including responsibility for security on the showground for three years in the early 1980s, membership of the Finance Committee from 1988 and as deputy show director from 1990, he offered imaginative, resolute and stylish leadership in discharging his duties as show director from 1994 to the present day.

A supportive media   One outside agency, the media, played a considerable part in promoting the Society and its show at Llanelwedd. In the first place, it was vital in the advertising campaign for the show as it was, later, for the Winter Fair. Greater efficiency in this department came with the appointment by Michael Creighton Griffiths, the managing director of Creighton Griffiths Advertising, in 1974 of Rob Petersen to handle the Society's advertising account, thereby establishing a partnership between the Royal Welsh Agricultural

238

Society and the Petersen Agency that would be most successful down to the present time. Certainly the agency was instrumental in promoting the image and awareness of the Society in the eyes of the target audiences and in ensuring attendance of growing numbers of visitors, old and new, each year. The importance of new visitors was underlined in the 1993 attendance survey which highlighted 17 per cent or 39,000 of them at the show. Mounting publicity costs – rising from £6,000 in 1976 to £34,000 in 1994 – went mainly on television advertising, which had been employed for the first time in 1975.

The media also brought the events of the show itself into the homes of many families. Press coverage saw the return year upon year of journalists like Roland Brooks, Leslie Able and Sheila Coleman (all three for the *Western Mail*), David Lloyd, John Price, Charles Quant, Barry Alston and Robert Davies. Initial accommodation for journalists at Llanelwedd in an all-wood structure adjacent to the bandstand, if noisy, marked an improvement on the 'chicken shed' of the migratory shows. In the later years they enjoyed the comforts of the ground floor of the International Pavilion and were made to feel that they were valued by the Society. The *Western Mail* and *Liverpool Daily Post*, in particular, reached a wide readership and invaluable, too, were the summer agricultural supplements of the *Western Mail* coinciding with the opening of the show. Testimony of the Society's gratitude to the press – 112 press representatives reported to the office during the 1982 show and 137 in 1983 – is reflected in the bestowal of honorary life governorship upon a select few, including Roland Brooks, in 1992, who besides reporting on the Llanelwedd show from its beginning to 1991 had also served on the Society's Editorial and Publicity Committee, and mid-Wales journalist John Price, in 1998, who had covered thirty-eight Royal Welsh Shows, and David Lloyd, in 1999, who had started off at Margam in 1959.

Radio and television provided a wider coverage of Llanelwedd as the years passed. Looking back in 1976 over his sixteen years as BBC agricultural producer in Wales, David John observed how the BBC's policy had changed over the years from initially mounting just a couple of prestige programmes to the provision of blanket coverage by the mid-1970s. In so doing, he declared, the BBC in Wales, both radio and television, was to give the Royal Welsh massive attention compared to the sparse coverage given to the Royal Show at Stoneleigh on English transmitters. Radio coverage of the 1977 show was greater than ever before and extension of coverage has continued down to the present day. Chris Stuart's first Royal Welsh Show came in 1987, thereby providing a new avenue to *Good Morning Wales*, and that same crucial show, with Welsh farming poised at its crossroads, was observed, too, by agricultural presenter Gaina Morgan. Television was arguably best able to capture the spectacle and drama of the event. The BBC became more actively involved with the show by transmitting live television coverage direct from the showground from 1977 onwards. The commitment of the media was further demonstrated when HTV made a film of the Society in 1979 carrying the title *Beastly Time*. Facilitated by a new team specializing in coverage of events, BBC Wales increased its presence at the show in 1988, devoting more than fifteen hours to television coverage. Still more would follow: in 1989, BBC Wales provided over eighteen hours of programmes, not including the time devoted each evening by *Wales Today*. In 1989, too, live transmissions were networked for the first time to the whole of the United Kingdom on Monday and Wednesday afternoons. BBC Wales's television coverage of the 1992 show drew criticism from the Society, however; it was a major disappointment that there had been no 'live' coverage and disapproval was expressed about the urban bias and what was perceived to be a trivializing of the event. Misgivings about the coverage of the show and agricultural programmes in general were expressed by the Society, the Country Landowners' Association, the FUW, the NFU and Wales YFC to a high-powered delegation of BBC Wales at a meeting on the showground in January 1993. There it was confirmed by the BBC that the Rural Unit at Bangor would continue and would be responsible for the 1993 show output rather than the urban unit used in 1992. Unfortunately, there would be no 'live' coverage of the 1993 show, it was announced by the delegation, since it would be uneconomic to produce it in English only following S4C's decision not to transmit live coverage in Welsh. Although the BBC's television coverage was criticized in 1993 and 1994 on the grounds that there was no 'live' provision, that it compared badly with the coverage given the National Eisteddfod and that there was a tendency to dwell on the peripheral items at the expense of the agricultural content of the show, nothing but satisfaction was expressed at the BBC's television coverage of the 1995 event, even though there was still no live coverage. In these delicate dealings with the media in the early 1990s the Society was well served by its Editorial and Publicity Committee under the skilful chairmanship of Montgomeryshire hill farmer, John Vaughan, who would hold that office from 1987 to 2000.

Another 'outside' agency important in sustaining the Society in its mission is the religious service which occurs on the Sunday before the opening of the show. From its early days at Llanelwedd, a show service came to be held at St Mary's, Builth Wells, and the Society was indebted to Canon Elwyn John for the welcome he gave them over many years. Innovation came in 1995 with the holding of an extra service on the Sunday, late in the evening on the showground, along the lines of *Songs of Praise* which then, as in later years, received a big input from the feature county.

CHAPTER FOURTEEN

Wider Horizons

$\mathbf{O}$ne major development at Llanelwedd, the Winter Fair, was perceived as fulfilling the same function as the summer show itself in providing another shop window for Welsh livestock. With such a fine facility at Llanelwedd there was an understandable feeling that the Society should use it for more than one important event a year. Besides, the Society was aware of the achievements of Welsh producers in both the live and carcass sections of the London Smithfield Show in recent years and envisaged a Royal Welsh Winter Fair as providing a much-needed stepping stone for producers to show the best of their produce on home ground. As such, this echoed the motivation of the founding fathers in 1904 in instituting a national show for Wales. After welcoming the success of the first Winter Fair, David Walters

91. A packed ringside at the 1991 Winter Fair.

rightly predicted that it would play a helpful role in promoting the Welsh farming industry in the years to come: 'This is a good start to what will without doubt become an important annual event in the farming calendar of Wales.' From the very first one-day fair held at Llanelwedd on 27 November 1990, the event, featuring Welsh prime livestock, food and crafts, established itself as a top-class occasion in the winter calendar of prime-stock shows. Farmers from Wales and beyond gave this first new major farming event in Wales for many years their full support despite the difficulties confronting the industry. The Winter Fair mirrored the success of the show itself in terms of the competitive spirit and keen interest displayed among livestock producers. Nevertheless there were real differences between the two shows, the principal one being the commercial function of the Winter Fair which was, indeed, very much a trade show. At the afternoon auction, winners appeared in the sale ring as at the Royal Smithfield and were sold to the highest bidders.

Benefiting initially from the sponsorship of the Midland Bank and Bibby Agriculture and, later, from Midland Agriculture and Dalgety Agriculture, the attraction of the event was reflected in the levels of attendance over the years (9,507 in 1995 stood as a record until it was overtaken in 2001 when 11,417 attended), rising numbers of livestock entries from farms throughout Wales and the borders, busy and growing numbers of trade stands and high bids for the supreme cattle champion. As in all such events, innovations and improvements occurred over the years; as early as 1992 the inclusion of the new Food Hall organized by Taste of Wales allowed for further expansion and with the 1999 Winter Fair centred around the recently opened Exhibition Centre the event was given a new dimension as well as more space and flexibility. Indeed, without the centre the eleventh fair in 2000 would have been a washout, given the deplorable weather. Such was the success of the venture that December 2002 saw the introduction of the first two-day Winter Fair.

Much of the new venture's success could be attributed to Verney Pugh, the director, and Derick Hanks, chairman of the committee, who took over as director following Mr Pugh's retirement after the 1996 event when he was joined by George Hughes as the new chairman. Other key personalities were (before his death in 1994) Winston Bowen, one of the instigators of the Winter Fair, and Patrick Tantrum, the chief cattle steward. With the institution of the Winter Fair a real landmark in the Society's history of progress had been reached. Amid the difficulties besetting all sectors of the meat trade, particularly in Wales, it helped to heighten public awareness of these challenges, provided Welsh farming with a platform where it could display the best of agricultural production and became a big commercial attraction to buyers. It certainly helped with confidence building within the industry and acted as a catalyst for better prices for Welsh livestock.

The years at Llanelwedd saw the Society's horizons ever widening, so that it came to represent much more than an annual show and, belatedly, a Winter Fair. Increasing numbers of out-of-show events and competitions were inaugurated, numerous conferences and seminars were arranged and the Society was consulted far more by institutions like the then Welsh Office and the Ministry of Agriculture, Fisheries and Food over matters relating to the farming industry. Significantly, it was only with the onset of profit-making shows from the mid-1970s that the Society came to have the resources and site facilities to further its aims and objectives as set out in its articles of association.

Although the use of the showground for out-of-show activities was rare in the early years at Llanelwedd, things began to improve from 1973 when the newly established Welsh Kennel Club's show, the Royal Welsh One Day Horse event – supported by the British Horse Society – and the regional trials for the Prince Philip Pony Championships were all held on the site and became annual features. From 1980 the Welsh Trade Fair (Wales Craft Council) was held in the South Glamorgan Exhibition Hall for the first time, migrating there from its earlier venue at the Hotel Metropole, Llandrindod Wells. Such was progress that, by 1982, the list of out-of-show events was lengthy and varied, that year in itself seeing a marked increase in demand from local auctioneers for holding various sales, including machinery, stock and furniture. In terms of actual numbers, 1983 saw more than sixty events held on the showground, involving 120 days' use of different facilities and generating a rental revenue of £30,000. These included sizeable bookings such as accommodating 2,500–3,000 people during the Wales Bible Week; the Welsh Kennel Club Championship Show; the six-day Auto Enduro event in October; the Welsh Craft Fair; the Mid Wales Business to Business Exhibition; and many sales of stock, machinery, antiques and furniture. Significant among the latter was the National Sheep Association Ram Sale; instituted in 1978 and the brainchild of Verney Pugh, the September ram sale for Wales and the border region held at Llanelwedd later became the biggest catalogued ram sale in the UK. Such was its outstanding success that the Wales and Border Ram Sales Committee decided to hold a new early ram sale at Llanelwedd for early lambing flocks in 1989. The year 1985 saw the first Welsh Black cattle sale established on the showground, while November 1991 witnessed the first autumn multi-breed show and sale of pedigree beef cattle on the site. On top of its hosting the Welsh Pony Championship from 1973 onwards, the showground in August 1990 was the venue for the International Pony Team Championships, so helping to push Wales and its ponies to the forefront. That prestigious event was to come to Llanelwedd through the determination of Rhandirmwyn farmer Will Jones, chairman of the Welsh committee of the parent British Show Pony Society.

We have seen earlier in this volume that extra-mural, out-of-show competitions were undertaken by the Society before 1963 in order to carry out its primary object to promote agriculture, horticulture and forestry, particularly in Wales and Monmouthshire, and to advance scientific research in connection with agriculture and forestry. These competitions included the Woodlands and Plantations Competition (from 1950), the Inter-County Hill Flock Competition (from 1955), the Replanning of a Farmstead Competition (from 1956), the Society's Silver Medal Competition for new machines or implements likely to facilitate farming on the Welsh uplands along with, from 1957, the D. Alban Davies Trophy and, from 1961, the Farm Machinery Maintenance Competition and the D. Walters Davies Trophy. From 1962 the All-Wales Ploughing and Hedging Match – in being since 1959 – was organized by the Royal Welsh Agricultural Society's secretariat. And the Society's premier award made annually from 1957, the Sir Bryner Jones Award, continues to the present day.

Certain of these competitions were later modified and new ones introduced once the move to Llanelwedd had taken place. The year 1963 thus saw a new competition announced called the Farm Buildings and Works Competition, a modification of that of 1956, and also another new one styled the Society's Silver Medal Award for New or Adaptations of Machines or Implements, which was a follow-up of the earlier Silver Medal competition for new machines, by offering a cash award to the *original designer* of the winning equipment. Organized by the Federation of Welsh Grassland Societies in conjunction with the Royal Welsh Agricultural Society, the Grassland Farming Competition was introduced in 1971 to promote the management and utilization of grass and its contribution to the farm economy. In 1983 came the introduction of the Meuric Rees Countryside Caretakers Award which recognized notable achievements by farmers who continued to care for the landscape and wildlife whilst producing food, although this would be dropped from 1992 because the Countryside Council for Wales (joint organizer with the Royal Welsh Agricultural Society) no longer deemed it appropriate, given that the approach to conservation had changed considerably. However, the two bodies jointly organized the Agri-environment Award for 1998–2002. At the same time the long-running Clean and Tidy Farm Competition staged in association with Keep Wales Tidy Campaign was discontinued from 1998, whilst specific themes were introduced in the Farm Buildings and Works Competition in 1999. The first Royal Welsh Tree-Felling Competition, with the aim of promoting safety aspects of chainsaw usage, was held in 1989 and took place annually in June at alternating woodland venues in north and south Wales.

Perhaps the most thoroughgoing modification in any of the competitions came in respect of machinery. After several unsuccessful attempts to make the old Silver Medal Award scheme work, in 1989 the Society determined to start anew by dropping all reference to silver and gold and to replace them by the D. Alban Davies Trophy Award – presented to the Society in 1957 – for the 'machine, implement or device likely to be of most benefit to Welsh upland farming'. A novel feature of this fresh approach to the machinery competition was the stipulation that no prior entry was required, the judges henceforth visiting every appropriate stand to select the piece of equipment which they considered to be of most benefit to Welsh upland farmers. Further modification occurred once again in 1998 when the award was made for 'devices likely to be of benefit to all Welsh farming'.

As an ad hoc committee of the Society was to confirm in 1971, the impact on farming within these specialized branches was very considerable, for the visitations of the appointed judges as well as their reports issued subsequently were eagerly sought, not only by the farmers and estates concerned, but also by those who were most closely linked with the particular subject. Where possible, too, demonstrations were held to follow up the awards and this served to promote good public relations with farmers, particularly those distant from the permanent site. Furthermore, in almost every one of the competitions, there was immediate contact with a large membership associated with the subject and, indeed, several organizations interested in such a competition, for example, the Welsh Mountain Sheep Society, the Livestock Department of the Ministry and the University Colleges with the Hill Flock Competition. Winning these competitions bestowed recognition and status; no others achieved such distinction as David and Gwen Davies who, farming Gwarffynnon, Silian, near Lampeter, carried off all the major Royal Welsh Agricultural Society awards – apart from student prizes – between 1984 and the end of the century.

The organization of conferences to discuss vital issues concerning Welsh agriculture had been started by the Society in 1954, and that same priority held good throughout the Llanelwedd years. One such conference convened at Llandrindod Wells in November 1969 was in response to the deep concern felt by Welsh sheep farmers about the trends in their industry, faced as they were with a fast-falling sheep population. The exchange of views among flock masters, farmers, auctioneers, butchers and Ministry advisers – notably the impressive Emrys Jones – on the future of the sheep industry in Wales was described by agricultural correspondent Charles Quant of the *Liverpool Daily Post* as a 'unique and vital conference' and he paid tribute to the 'statemanship' of the Royal Welsh Agricultural Society in arranging this gathering. From the close of the 1980s the Society also became much involved in the organization of two major conferences a year, namely, the Welsh Agricultural Outlook

Conference (from 1988), organized and sponsored by the Society, the Agricultural Development Advisory Service (ADAS), Wales Farm Management Association and Midland Bank; and the Welsh Farming Conference (from 1987), jointly organized by the Welsh Agricultural College and the Royal Welsh Agricultural Society and sponsored by Midland Bank. The latter conference was hosted by the Royal Welsh Show's feature county and held on the first Wednesday evening in November. Each conference reflected the farming of that area; for example 'Milk Marketing – a New Era' was the theme of the Welsh Farming Conference held at Haverfordwest on 7 November 1990. However, when the Welsh Institute of Rural Studies and the main sponsor, Midland Bank, decided that the event had run its time, the Welsh Farming Conference ceased to be held after 1997. One-off conferences were also sponsored by the Society, for instance, the RWAS–Lloyds Bank Milk Conference at Llanelwedd in November 1993.

As a crucial part of its mission to promote Welsh agriculture the Society placed great store, too, on organizing demonstrations and open days either on the showground itself or at various venues throughout Wales, often in co-operation with other bodies like the National Agricultural Advisory Service and ADAS. Moreover, since many farmers in outlying areas from Llanelwedd – who increasingly from the 1960s farmed one-man units – simply could not spare the time to visit the show, these practical demonstrations at sites distant from the showground served to give the members and other farmers a sense of belonging to the Society. The year 1973 saw a special grassland demonstration at Gelli-aur, Carmarthenshire, jointly organized by the Gelli-aur authorities, the Royal Welsh Agricultural Society and ADAS, which attracted an attendance of over 3,000. Another successful demonstration came with a joint open day with the Royal Agricultural Society of England at Cwm Whitton and Lower Nantygroes (Radnorshire) in May 1976, through the hospitality respectively of Mr Pugh and Mr Morgan and their families. Remarkably successful and long recalled was the first two-day national event devoted exclusively to Britain's hill farmers staged on the Rhiwlas estate, Bala, by permission of the owner, Robin Price, and one of his tenants, Gwynn Lloyd Jones, on 4–5 June 1986. Organized by the Welsh region of ADAS and the Society in association with *Farmers Weekly* and the National Sheep Association, the event was comprehensive in its coverage, and it certainly helped persuade many that the Society had a useful role to play in organizing national agricultural events outside the show. As noted, too, demonstrations sometimes came in the wake of prize-winning schemes awarded by the Society in its different competitions, for instance, those held on 1 December 1970 on the farms of Tynewydd, Carrog, and Tynllechwedd, Gwyddelwern, as the schemes placed first and third in the Farm Buildings and Works Competition.

It was the continued policy of the Society to assist other organizations sharing similar aims and objectives. In this respect, the Society's secretariat had done important work in assisting the breed societies before 1963. However, just before the move to the permanent site the secretaryship of the Welsh Halfbred Sheep Breeders Association was transferred from the Royal Welsh Agricultural Society to the Welsh Agricultural Organisation Society, Aberystwyth (founded in 1922), a transfer rendered necessary because the Royal Welsh had decided to be registered under the Charities Act 1960 which meant it could no longer provide secretarial facilities to commercial enterprises. The following year also saw the Welsh Pony and Cob Society undergo what was claimed to be a 'friendly separation' from the Royal Welsh secretariat and move back to Aberystwyth, a decision that drew criticism from Austin Jenkins, who viewed it as frustrating the Royal Welsh's aim of developing Llanelwedd as the agricultural centre of Wales. If the Welsh Pig, the Welsh Halfbred Sheep and the Welsh Pony and Cob Societies had thus severed their ties by 1964, the Royal Welsh nevertheless continued performing secretarial work for the Welsh Mountain Sheep Society, the Welsh Ploughing Association and the Livestock Export Committee for Wales. As before 1963, the cost of furnishing these facilities would continue to cause concern, so much so that, by the end of the 1960s, the Society sought to persuade the Welsh Mountain Sheep Society to pay more for services rendered.

An ever more important role for the Society as an independent body in Wales was its role in framing authoritative views in response to industry consultation documents referred to it by the Ministry of Agriculture, the Welsh Office and environment agencies. After the foot and mouth outbreak of 1968 an ad hoc panel, chaired by the Society's chief veterinary officer F. V. John, submitted a memorandum on the disease to the Northumberland Committee, and the Society derived satisfaction from the fact that the views it expressed coincided almost identically with the main recommendations of the committee. Around this time, too, specialist panels of the Society also prepared invited responses to the Ministry of Agriculture on the subjects of 'Sire Licensing of Bulls' and 'Evidence concerning the Vetinerary Profession'. By the 1990s the Society was being consulted far more than ever before and in preparing its reports it was well served in particular by Bill Ratcliffe, Robin Gill and Cyril Davies.

While careful to protect its standing and role as an independent, non-political body, the dire conditions facing Welsh farmers in the late 1990s saw Council becoming more strident in December 1997 in its demand for government action to deal with the BSE crisis. In so doing it was supporting similar demands made by the farming unions and other organizations in Wales. A determined Meuric Rees, chairman of the Council, stated:

The Royal Welsh is not a political lobby as such but we want our supporters and members to know that we stand fully behind them whether they are demonstrating to bring either the attention of the public to what is going on or whether they are the leaders of our industry who are knocking on the doors of the powers that be to find the solution.

Adopting a similar resolute stance, chairman of the Board, Dr Emrys Evans, voiced his opinion at the 2000 annual general meeting that the Society had an obligation to register its disquiet at the plight of the industry and its deep concern about the future. Such a proactive approach, he informed members present, was mirrored in the endless correspondence with members of the government and of the National Assembly for Wales and growing involvement in events and seminars organized by the Assembly and other relevant agencies. One issue among others listed by Dr Evans that was supported by the Board was a campaign to prevent small abattoirs from being forced out of business as a result of the draconian charges for meat hygiene service inspections.

For much of its history the Society had been aware of the *Journal*'s importance as a way of keeping members in touch with its activities and of imparting useful information on matters relating to Welsh farming. Stormy times, however, persisted from 1963 down to the year 1979. Mindful of the £921-cost of publishing and distributing it in 1962, some members of the Society were in favour of discontinuing publication in 1964 when faced with an adverse financial position. Nor, seemingly, were many much bothered about its fate. However, Dr Richard Phillips, supported by Professor White and the Honourable Islwyn Davies, prevailed on the Finance Committee to open its purse strings. A change in the format of the *Journal* occurred in 1965 in response to the sudden increase in membership. The title was changed to the *Royal Welsh Journal* and three slim issues appeared each year, in January, June and October respectively. No longer would articles relating to historical and scientific aspects of Welsh agriculture appear, and doubtless this was partly because information on up-to-date farming techniques could now be gained from the national farming press. Instead, the emphasis was placed on projecting the image of the society to the community at large in order to attract people to the show. Such was the continuing annual financial deficit incurred by the Society, however, that in November 1966 Lieutenant Colonel FitzHugh stated firmly that no money would be made available for publication the following year, and it would therefore survive by publishing just two issues – the annual and the show edition – by means of the

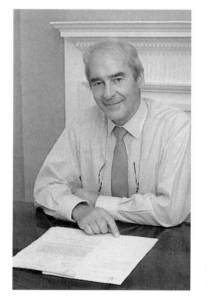

92. John Kendall, public relations consultant for the RWAS from the 1970s to the present day.

Lady Roberts grant of £250 a year. These tight financial limitations persisted for some time to come and it was only in 1979 that the clouds lifted when the new-style spring issue of the *Royal Welsh Journal* appeared. In the light of the favourable reception given by members to this more substantial publication, the Board agreed that the new style be extended to the show publication as well. So attractive was the special show issue of the *Journal* for 1987 to mark the first twenty-five years at Llanelwedd that recommendations were made that the Society should produce one good quality *Journal* each year rather than the spring and show issues. Financial exigencies would once again see its size reduced in 1994, but its quality was nevertheless maintained under the capable editorship of John Kendall. Its attractive glossy format, lively presentation of the whole range of the Society's activities and personalities, lavish provision of photographs and its blend of English- and Welsh-language items, ensured its success among its members as a living representation of their organization.

Looking Back and Forward

Few predictions can have been wider of the mark than that of John Gibson when he wrote in February 1904: 'This show for the whole universe will . . . come to nothing.' In its centenary year the Society has much to celebrate. Not least of its achievements is the fact that it has successfully coped with the recurring problem of the most suitable location for the annual show, a challenge rendered particularly difficult because of the configuration of Wales. After only a few years at Aberystwyth, the Society, responding to a deeply felt sentiment that to be truly 'national' the show should migrate between centres in the north and south, cut adrift from that seaside town in 1910. Those peripatetic shows – thirty-seven in all – were perceived down to the early 1950s as the most satisfactory way of serving the farming community of Wales, but from that time onwards mounting costs of staging the show and the rising expectations of visitors over basic facilities gave the society little choice but to settle on a permanent site. Besides, despite the huge efforts to make the show a national event, rather too many people viewed it as just a north Wales show one year and a south Wales one the next, an attitude that translated into disappointing membership subscriptions and show entries. Increased membership, it was believed, could only be achieved if the show moved to a permanent site.

Although Aberystwyth was the first choice and other mid-Wales towns, too, were considered, the decision to favour Llanelwedd was a wise one. Despite the strong feeling of many that Aberystwyth was, historically, the show's true 'home', the seaside town, apart from lacking a suitable site, was not central enough, and it is inconceivable that those many English competitors who came to Llanelwedd would have journeyed further west to Aberystwyth. As a permanent site, Llanelwedd's location was as good as anywhere else in Wales; it was reasonably central for the rest of Wales and accessible from over the border. Arguably, Aberystwyth would have been the best centre for a permanent National Eisteddfod site. Although, of course, the Eisteddfod would remain migratory, it is interesting to note that in the late 1950s its representatives, including Principal Thomas Parry of the University College of Wales, Aberystwyth, met those of the Society at the Montgomeryshire home of the Honourable Islwyn Davies, to discuss the possibility of joining forces in seeking a permanent site. Under discussion was the idea of forming a company to buy a couple of farms, one in the south between Port Talbot and Bridgend, the other in the north along the coastal strip, which would be developed to stage the Eisteddfod and the show, the latter being held one year in the north while the Eisteddfod would be staged in the south, the position turning round the following year.

Although the Society's early years at Llanelwedd were accompanied by discussion about the seeming lack of appeal of the new site and the possible advantage that might accrue if a new showground was chosen, with the increasing outlay of capital at Llanelwedd a transfer became less and less of an option. What needed doing urgently was to convince people that the Society was there to stay, and no more powerful statement was made on this score than the building of the grandstand ready for the 1980 show. After its erection, people from the north came to accept the fact of the permanent site. So successful did the permanent showground become in attracting people that Lord Gibson-Watt recalled with amusement how during his term as chairman of the Council he was to be asked by a visitor to one crowded show: 'Why don't you make one-way traffic here?'

Pride can be taken, too, in the way that the Society built itself into a prosperous organization, even though a serious setback was inflicted by the foot and mouth crisis of 2001. This achievement came about despite adverse factors like the initial jealousies of local agricultural societies, poor weather ruining shows, mounting show expenditure largely driven by increasing erection costs, and sluggish, sporadic membership, and it owed much to the Society's willingness during difficult financial years to undergo rigid economies in every possible direction. Of particular importance in securing the necessary funds for developing Llanelwedd was the unique system of feature counties, the envy indeed of other large agricultural societies within the United Kingdom. Also crucial in accounting for the Society's progress in its Llanelwedd years was that as a charity it paid no tax. Another important influence in promoting its well-being throughout was the dedicated, mostly voluntary service of officials and stewards. And here it is proper to recognize the crucial role that the old landed families of Wales played in setting up the Society in 1904 and the service that many of their descendants have given over the first hundred years of its history. Their efforts in 1904 were to belie some of the criticisms levelled against them at the time by certain sectors of Welsh society that they were a harmful influence on Welsh rural development. A prominent member of the Society, Dr Richard Phillips, was to write in an essay entitled 'From Aberystwyth to Machynlleth', published in 1954: 'They played their part on the County Councils and took the leadership in every new movement, counting it a duty and a responsibility to care for the welfare of their tenants. To them we are principally indebted for the foundation of this Society, and the descendants of many of them continue to work loyally for it.' In addition, the links maintained with the counties by means of county meetings from 1924, superseded from 1961 by county advisory committees, the strong support forthcoming in the early years from the University Colleges of Wales, particularly Aberystwyth and Bangor, the professionalism of its small permanent staff, royal patronage and show visits, the involvement of the Young Farmers' movement, and strong backing from the media all helped to ensure the Society's growth. Undoubtedly part of its success, too, stemmed from the ability of its organizers after 1963 to strike the right balance between preserving the essential agricultural flavour of the show and opening it up to the general public, whether non-farming groups in the countryside or town dwellers. In like fashion, it reached a sensible balance between English- and Welsh-language provision.

Sustaining the Society's relevance and impact throughout its history has been its constant willingness to develop and innovate in order to keep abreast of the times and so be able to reflect and propagate improving farming methods and marketing techniques among the agricultural communities of Wales and the border counties. Many instances of this have been met in the foregoing pages, such as the promotion of dairying and forestry at its shows during the 1920s and 1930s, the regular introduction of new livestock classes, British and (later) foreign, and, from the 1960s non-pedigree cattle and carcass classes, the development of the Country Pursuits and Sports Area in the late 1970s, and promotion of Welsh food products from the mid-1980s. Out-of-show competitions, such as the Meuric Rees Countryside Caretakers Award instituted in 1983, similarly reflected new fashions.

While competitions in all sections of the show and out-of-show ones aimed to raise the standard of agriculture, horticulture and forestry, particularly in Wales, the Society was especially concerned to improve the quality of the native livestock. Not only did it inaugurate Inter-County Breed Competitions from the 1920s which would continue to 1939, but from the 1920s down to the late century it achieved much through its support of the Welsh breed societies. A similar concern for peculiarly Welsh matters was seen in the way it gave encouragement from the 1950s to the invention of farm machinery best suited to the difficult Welsh upland terrain.

As well as acting as a shop window or a showcase for Welsh agriculture and its ancillary industries and crafts and as a place for farmers to do business and to exhibit in order to promote their livestock, the educational side of the annual show was a vital means of informing the farmer about new and fast-growing scientific – mechanical and chemical – improvements in agriculture. That priority, which the Society has throughout placed on its educational role, was likewise discharged by its organization of conferences and demonstrations and by its support of agricultural courses in the field of higher education within Wales. In the pre-1963 years especially, the *Journal* was heavily geared towards disseminating information about the latest scientific experiments and new techniques.

Moreover, the show throughout fulfilled an important social role in Welsh rural life, all the more so given the absence in Wales of venues such as race meetings or Badminton Horse Trials where English people for their part could catch up with friends. For a week each year the farming people of rural Wales and its border counties have assembled not just to compete but to relax and enjoy themselves, many looking upon their visit as their annual holiday. This social dimension of the show was enhanced by the provision of caravan parks adjacent to the showground from the late 1970s; farming families from north Wales, in particular, were to find this facility advantageous. In very difficult farming times, too, the show has served as a vital morale booster, allowing farmers and their families to escape their isolation for a fleeting moment, to share their anxieties and to be reassured about the value of their industry. Indeed, it was the realization of the stimulus which the holding of a national show undoubtedly provided that persuaded the Society to go ahead as normal with the Royal Welsh Show in 1932 in the midst of economic crisis engulfing the country, and, again, without doubt the show helped to lift spirits in the midst of the deep farming gloom of the late 1990s and beginning of the new millennium, not least the show in 2002.

Those difficult years for the Welsh farming industry from the 1980s to the present day have also seen the show become an important, natural venue for official and unofficial discussions on the problems besetting the industry. Similarly, as an independent body not wedded to any particular interest group the Society, increasingly so in the last two decades of the twentieth century, has been relied upon to respond to consultation documents referred to it by government and other agencies.

Although the showground has been developed into a site affording superb facilities which have enabled it to function as the agricultural and rural centre of Wales in all seasons – the prestigious ram sales (organized by the National Sheep Association Wales and Border Regions) and Winter Fair spring to mind – certain projects dreamt up to give a boost to the Llanelwedd venture did not reach fruition. Nothing came of the idea at the outset to establish a National Arboretum in the vicinity of Llanelwedd Hall; the envisaged road racing circuit and associated motor museum did not get under way at the start of the 1970s; a Welsh Rural Life Centre at Llanelwedd, discussed around the same time in conjunction with the Wales Tourist Board, the Council for Small Industries in Rural Areas (COSiRA) and the Welsh Folk Museum with the aim of presenting the rural heritage of Wales, did not materialize; nor did the dream of T. Mervyn Jones (president in 1974) to establish the showground as a permanent exhibition and information centre for Welsh tourism come to fruition. Years later, in the early 1990s, the possibility mooted by the Development Board for Rural Wales of developing the site as a major regional sporting venue was deemed too expensive an undertaking. Had the bid for £7 million from the Millennium Commission in 1995 towards the overall £14-million cost of developing the showground into a centre of excellence for rural Wales been successful, the site would have been much improved, but it was inexplicably turned down. Finally, Hywel Richards's recommendation at the start of the new millennium – echoing an earlier aspiration – that a National Rural Life Museum be set up on the showground was rejected by the Board.

If there may have been some exaggeration in Lord Daresbury's pro-nouncement in 1927 – as honorary show director of the Royal Agricultural Society of England – that 'it is not too much to say that the high standard obtained among farmers in Britain today is due more to national and local shows than to all other agencies combined', full recognition of the Society's work in its early years was given by the *Western Mail* in July 1931:

The Society has done and is doing magnificent work for agriculture, especially in the department of stock-raising, which is the predominant agricultural interest of the Principality, and a moment's thought would convince anyone of the fact that these annual exhibitions are of vital necessity in raising the standards of stock-breeding and husbandry. This is a matter of special importance in Wales because the Royal Welsh Agricultural Society is in a special degree, and in an indispensable manner, the guardian of the national breeds of farm-stock as well as the patron of the celebrated border breeds.

What the show did in those days and has done ever since is, by bringing together the best, to provide an incentive to progress and a measure of progress already made.

While many readers will appreciate the show's enormous success in advancing from being just the highlight of the Welsh farming calendar to becoming by the 1980s one of the three major agricultural events in the United Kingdom – and at the same time retaining its friendliness and Welsh flavour – few will be aware of the impact of the Society and show upon the economy of rural Wales. An investigation of the economic impact of the showground over the year September 1992 to August 1993 carried out by the Department of Agriculture, University of Wales, Aberystwyth, estimated that gross expenditure amounting to £24.1 million was generated by showground activities, including capital developments, during that period. Some two-thirds of this, approximately £15.8 million, arose from the 1993 show, while miscellaneous out-of-show activities, including the Winter Fair, accounted for a further £5.7 million. The remaining £2.5 million of total expenditure was accounted for by the operation of permanent buildings on the site, the other administration and organizational costs and, finally, additional purchases stemming from capital developments.

Spending on the showground accounted for £13.9 million (58 per cent) of the gross volume of expenditures associated with showground activity, and over three-quarters of this occurred during the show period itself. Total expenditure outside the showground amounted to over £10 million, just under half of which took place within a 25-mile radius of Builth Wells and within Wales. The economic consequences of the spending were estimated to create local income to the value of £2.04 million, representing an average of £79 per household in Breconshire and Radnorshire, sufficient to support 147 direct and secondary jobs in Builth Wells itself and the surrounding area. Growth in showground activities since 1992–3 means that at the present day perhaps in excess of £35 million a year gross expenditure is being generated.

Not least among the mid-Wales beneficiaries of the Llanelwedd permanent site were the hotels, guest houses and public houses, the boost given their businesses spilling outwards from the town of Builth Wells itself to neighbouring centres like Llandrindod Wells and even Presteigne, eighteen miles away. Such was the pressure for accommodation from the outset that one local hotel manager before the 1964 show claimed that a man had tried to persuade him to let him sleep in the bath tub for the night!

Standing alongside the National Eisteddfod of Wales as one of the two major Welsh institutions, the Royal Welsh Agricultural Society has over the years succeeded in bringing north and south Wales together, in bridging the gap between town and country, and in harmoniously incorporating Welsh speakers and non-Welsh speakers within the same movement. This imaginative embrace went far towards explaining the *Western Mail*'s headline on 20 July 1999: 'Sioe Frenhinol Cymru yn mynd o nerth i nerth' (The Royal Welsh Show is going from strength to strength). Despite the tribulations suffered during and since the foot and mouth outbreak of 2001, the complete recovery in livestock entries at the 2003 show (see Appendix Four) as well as the attendance of 213,538 – higher than that at Stoneleigh – meant that the Society completed its first hundred years on a high note. It can look forward with confidence to a new century of growth and development. Of course, certain problems will have to be resolved, in particular the two highlighted in the recent research findings of Jane Ricketts Hein for her M.Sc. degree dissertation. Firstly, it will be important not to dilute further the agricultural content of the show otherwise trade-stand holders may well turn their backs on it in preference for specialist agricultural events. And, secondly, past and current involvement of the Young Farmers' Clubs in the show notwithstanding, there will be a need for younger farmers to play a bigger part than hitherto in the management of the show and the Society. These apart, it will also be helpful and appropriate if the very recent trend towards involving more women in the active running of the Society – notably prompted by Teleri Bevan in her spirited address when opening the show in 1999 – is given greater emphasis. Above all else, the Society will need to consider how it can best serve the far from buoyant Welsh rural scene – encompassing agriculture, forestry and tourism – and assist in the regeneration of its villages. Its feature-county arrangement and its customary readiness to adapt to change and to develop in new directions well equip the Society to deliver its challenging agenda in the years that lie ahead.

Appendices & Bibliographical Note

Comparative entries 1911–1939

SECTION	1911	1912	1913	1914	1922	1923	1924	1925	1926
Traders stands	137	81	64	78	127	118	130	128	137
Welsh ponies & cobs	102	86	55	81	81	88	78	113	50
Welsh ponies & cobs – local	–	33	–	–	–	13	–	–	5
Riding cobs & ponies	–	–	–	–	12	44	34	57	18
Hackneys	24	13	38	26	37	29	15	16	14
Hunters	34	20	8	39	66	86	66	82	41
Hunters – local	–	–	–	–	–	43	10	31	–
Shires	148	51	41	59	60	92	49	50	43
Shires – local	–	23	–	–	11	87	35	29	48
Colliery horses	–	2	3	18	13	–	10	22	–
Harness classes	24	50	13	28	35	32	37	44	35
Jumping	–	–	–	–	–	–	–	–	–
Welsh Black cattle	95	50	120	42	126	125	125	142	260
Shorthorn cattle	49	34	27	44	87	41	83	95	65
Hereford cattle	81	61	42	45	56	39	50	37	19
British Friesian	–	–	–	–	40	18	26	24	25
Park cattle	–	–	–	–	–	–	–	7	–
Local cattle	–	–	–	–	103	11	58	71	49
Dairy cattle	1	–	2	2	–	–	–	–	–
Dairy cattle – local	6	15	–	–	–	–	–	–	–
Recorded cattle	–	–	–	–	–	–	–	–	–
Premium cattle	–	–	–	–	–	–	–	–	–
Milking trials	–	–	–	–	–	–	–	–	–
Other cattle	–	–	–	–	–	–	–	–	–
Aberdeen Angus	–	–	–	–	–	–	23	–	–
Goats	–	–	–	–	–	–	–	–	–
Goats – local	–	–	–	–	–	–	–	–	–
Welsh Mountain sheep	102	39	108	62	112	63	81	58	88
Black Welsh Mountain sheep	–	–	–	–	13	20	21	23	13
Welsh wool	–	–	–	–	–	–	–	–	–
Ryeland sheep	6	16	7	12	8	6	13	16	–
Kerry Hill sheep	69	–	36	42	72	99	67	92	67
Shropshire sheep	50	–	21	17	16	14	–	26	–
Shropshire sheep – local	–	–	–	–	–	–	–	–	–
Southdown sheep	–	–	–	–	–	–	–	–	23
Suffolk sheep	–	–	–	–	–	–	8	9	–
Wiltshire sheep	–	–	–	–	–	–	–	–	18
Local sheep	–	–	–	–	83	–	–	–	62
Clun Forest sheep	–	–	–	–	–	–	–	–	–
Other sheep	–	–	–	–	–	–	–	–	–
Pigs	19	–	16	20	77	105	132	193	104
Pigs – local	–	–	–	–	11	1	–	26	–
Leaping	18	26	5	20	21	53	33	53	60
Trotting	–	6	5	16	–	–	–	–	–
Tradesmen's turnouts	–	–	–	–	–	–	–	–	–
Dairy produce	40	56	16	46	216	130	167	85	105
Honey	11	26	39	37	21	55	33	102	83
Bread	–	–	–	–	250	85	86	32	37
Buttermaking	24	34	12	84	27	66	70	75	75
Cider	–	–	–	18	–	–	–	–	–
Poultry (rabbits, pigeons)	–	–	–	–	788	498	863	544	637
Dogs	–	–	–	–	1273	940	1270	478	764
Hounds (couples)	–	–	–	–	90	93	100	74	–
Forestry	–	–	–	–	–	35	108	123	112
Horticulture	–	–	–	–	257	–	167	337	360
TOTAL ENTRIES	972	722	678	836	4189	3123	4105	3321	3360

1927	1928	1929	1930	1931	1932	1933	1934	1935	1936	1937	1939
115	164	172	124	141	139	131	160	154	161	161	142
92	77	66	61	49	43	53	57	49	45	39	45
16	14	11	18	6	13	51	28	8	7	5	2
37	49	58	35	32	41	48	53	51	111	85	74
27	18	18	26	14	11	8	7	7	12	9	9
41	91	48	29	72	47	44	71	81	88	72	40
21	16	9	—	9	2	1	—	94	—	50	—
35	55	36	64	35	48	52	76	42	71	53	66
26	20	10	63	18	26	35	21	64	57	29	30
23	—	13	—	10	—	—	—	4	—	9	—
71	37	88	36	46	29	30	34	34	42	46	31
63	61	77	63	107	120	113	185	180	215	68	99
82	101	66	109	67	55	76	80	73	97	40	83
75	79	74	94	86	99	107	105	100	93	121	72
27	26	38	25	26	54	31	36	38	29	61	19
18	29	22	31	32	21	—	22	21	31	24	22
—	—	—	—	—	—	—	—	—	—	—	—
—	79	37	39	31	53	63	50	102	81	81	58
—	—	—	—	—	—	—	—	—	—	—	—
—	—	20	28	15	16	8	16	13	12	8	8
12	7	8	15	11	11	13	5	10	8	4	6
—	—	12	16	13	15	9	9	12	9	9	7
—	—	—	—	—	—	—	15	13	18	27	12
—	—	—	—	—	—	—	—	—	—	—	—
—	26	18	13	27	41	37	47	44	51	34	49
—	—	—	—	—	—	—	—	—	—	—	—
74	81	54	70	54	50	68	83	45	82	65	83
	17	18	9	11	9	17	9	9	16	15	18
—	—	—	—	—	6	13	9	14	13	8	15
17	19	22	21	22	—	—	—	—	—	22	14
89	101	67	61	53	58	56	35	31	34	26	32
—	—	—	—	—	—	—	—	—	—	—	—
—	—	—	—	—	—	—	—	—	—	—	—
—	—	—	—	—	—	—	—	—	—	—	—
—	—	—	—	—	—	—	—	—	—	—	—
—	102	60	54	16	39	47	56	55	76	46	47
—	18	18	19	18	40	45	36	31	35	32	28
18	12	14	12	10	10	8	44	—	34	—	14
84	125	84	79	72	80	82	110	129	120	92	81
—	11	11	11	5	7	17	15	58	14	12	7
—	—	—	—	—	—	—	—	—	—	—	—
—	—	—	—	—	—	8	4	4	—	—	—
96	198	109	140	114	152	148	171	150	101	128	186
36	33	111	90	81	57	141	205	122	114	87	
38	22	11	24	45	42	50	95	51	82	252	86
70	52	151	81	84	87	80	93	80	85	118	57
—	—	—	—	—	—	—	—	—	—	—	—
737	464	601	561	742	520	676	785	678	770	677	559
883	809	—	—	—	—	—	—	—	—	—	—
87	55	70½	—	46½	—	—	—	—	—	—	—
106	138	167	217	181	—	—	—	—	28	56	50
637	284	54	313	186	129	429	390	819	688	134	365
3777	3490	2523½	2651	2589½	2179	2829	3226	3480	3620	2783	2610

Comparative statement of entries for the years 1947, and 1949–1962

SECTION	1947 Carmarthen	1949 Swansea	1950 Abergele	1951 Llanelwedd	1952 Caernarfon	1953 Cardiff	1954 Machynlleth
Traders' stands	115	201	236	244	254	289	245
Welsh ponies & cobs	94	122	101	129	108	114	133
Welsh ponies & cobs – local	20	55	38	36	27	35	48
Riding cobs & ponies	55	60	70	82	36	74	47
Riding cobs & ponies – local	–	45	27	14	21	36	11
Hackneys	3	–	–	–	–	–	–
Hunters	70	37	42	70	72	49	79
Hunters – local	73	34	13	4	–	20	6
Shires	55	38	48	51	29	39	48
Shires – local	44	20	50	4	39	14	13
Colliery horses	–	30	–	–	–	31	–
Harness horses	21	11	50	56	38	39	30
Harness horses – local	–	10	1	–	–	–	–
Jumping	62	104	241	192	191	236	228
Welsh Black cattle	71	77	65	87	–	48	89
Shorthorn cattle	99	88	60	126	–	73	63
Hereford cattle	31	41	45	67	–	66	50
British Friesian cattle	15	28	41	71	–	49	72
Ayrshire cattle	52	52	92	90	–	53	79
Jerseys	–	–	–	–	–	–	–
Guernseys	–	–	–	–	–	–	–
Aberdeen Angus	–	–	–	–	–	–	–
Other cattle	–	–	31	63	–	54	76
Cattle – local	177	142	116	42	–	57	120
Recorded cattle	–	–	–	–	–	–	–
Milking trials	9	6	10	–	–	–	–
Welsh Mountain sheep	47	82	88	59	–	54	102
Black Welsh Mountain sheep	–	–	–	–	–	7	–
Wool classes	–	–	3	–	15	24	40
Ryeland sheep	6	6	4	33	–	23	26
Kerry Hill sheep	30	38	35	36	–	29	47
Clun Forest sheep	29	35	33	64	–	52	59
Suffolks	–	–	–	–	–	–	–
Border Leicester	–	–	–	–	–	–	–
Other sheep	–	–	14	52	–	94	99
Sheep – local	34	96	92	73	–	68	147
Pigs	61	56	95	165	–	201	237
Pigs – local	22	28	36	2	–	61	34
	1,295	1,567	1,777	1,912	815	1,989	2,228
Dairy products & preserves	–	–	96	113	180	190	300
Honey	–	206	161	246	249	215	283
Bread & confectionary	–	–	83	209	135	140	181
							no rabbits
Poultry, pigeons & rabbits	524	–	435	–	392	616	387
Forestry	25	32	33	25	38	21	27
Woodlands	–	–	14	26	19	37	58
Horticulture	306	227	290	22†	630 29★ }	304 35★	15 13★
Crafts	41	74	107	115	77	85	134
Sheepdog trials	37	86	58	86	92	69	89
TOTAL ENTRIES	2,228	2,192	3,054	2,754	2,656	3,701	3,715

† Trade stands only   ★ Horticultural traders only

Note: the cloven–hoofed classes had to be cancelled in 1952 due to outbreak of foot and mouth disease. The poultry, pigeon and rabbits classes had to be abandoned in 1949 owing to fowl pest.

1955 Haverfordwest	1956 Rhyl	1957 Aberystwyth	1958 Bangor	1959 Margam	1960 Welshpool	1961 Llandeilo	1962 Wrexham
272	267	250	288	295	304	326	321
115	138	131	130	136	146	219	228
62	– –	–	–	–	–	–	
62	130	128	115	106	165	123	121
64	–	–	–	–	–	–	–
83	108	61	47	47	84	145	120
66	–	–	–	–	–	–	–
31	77	38	39	32	52	22	51
–	–	–	–	–	–	–	–
–	–	–	–	7	–	9	–
29	39	29	21	27	25	19	11
–	–	–	–	–	–	–	
91	196	301	309	396	352	145	281
73	86	130	88	49	67	76	75
71	72	40	27	33	44	52	63
52	49	37	32	47	71	53	35
79	88	50	63	43	86	79	53
62	91	55	45	39	69	51	47
–	–	–	–	–	62	40	36
–	–	–	–	–	23	23	41
–	–	–	–	–	11	12	6
92	70	57	27	64	–	–	–
276	–	–	–	–	–	–	–
–	–	–	–	–	–	–	–
28	50	87	78	68	117	84	109
–	–	–	–	–	–	–	–
19	15	29	24	19	–	–	–
33	20	16	–	21	9	14	11
18	30	34	–	17	30	23	33
43	51	48	38	20	53	41	25
–	–	–	–	–	16	47	40
–	–	–	–	–	46	31	53
77	156	142	187	125	50	96	106
35	–	–	–	–	–	–	–
182	136	153	118	125	181	127	148
136	–	–	–	–	–	–	–
2,151	1,869	1,816	1,676	1,716	2,073	1,878	2,026
265	211	275	348	297	307	317	343
136	313	234	244	248	279	253	280
181	186	269	190	142	155	222	258
	no rabbits from 1956 through to 1962						
666	748	700	536	592	630	722	641
–	32	20	45	9	14	33	42
32	50	22	27	37	32	44	47
–	–	–	–	–	34	74	124
16★	35★	–	24★	21	17★	34★	39★
124	161	34	37	43	–	–	36
58	–	–	–	–	–	–	–
3,629	3,605	3,370	3,127	3,105	4,341	3,577	3,836

Summary of results for the thirty-nine years ended 31 December 2002

FEATURE COUNTY		ATTENDANCES	SUBSCRIPTIONS £	GATE RECEIPTS & ENTRY FEES £	OVERALL PROFIT £	DEFICIT £	EXPENDITURE ON PROPERTY £
1963 (3 days)	Radnor	42,427	7,659	32,586		14,421	68,682
1964 —	Brecon	65,348	12,452	41,913		9,380	14,179
1965 —	Anglesey	59,419	14,852	43,049		10,445	11,979
1966 —	Pembroke	64,530	13,840	43,759		12,196	10,196
1967 —	Cardigan	71,256	15,794	47,166	366		8,057
1968 —	Merioneth	60,163	17,100	41,777		8,723	7,193
1969 —	Glamorgan	72,840	15,530	47,776		2,296	4,228
1970 —	Caernarfon	55,117	15,082	41,556		8,076	20,806
1971 —	Monmouth	63,561	14,738	45,537		6,008	3,426
1972 —	Denbigh/Flint	67,337	14,508	53,285		4,913	8,972
1973 —	Carmarthen	77,024	18,709	81,356	6,673		81,825
1974 —	Montgomery/Powys	79,446	18,469	85,397		5,829	57,191
1975 —	South Glamorgan	90,036	19,518	109,484	9,440		81,251
1976 —	Radnor/Powys	105,026	26,675	159,373	38,615		136,526
1977 —	West Glamorgan	111,162	32,539	195,905	69,922		93,297
1978 —	Merioneth/Gwynedd	136,215	37,029	253,102	90,318		61,155
1979 —	Pembroke/Dyfed	141,327	42,953	320,314	115,183		240,522
1980 —	Brecknock/Powys	122,777	50,342	336,660	77,523		345,620
1981 (4 days)	Clwyd	155,606	55,497	403,072	89,715		87,170
1982 —	Mid Glamorgan	146,821	75,428	405,399	72,389		164,590
1983 —	Ceredigion/Dyfed	159,157	63,650	454,006	88,079		94,146
1984 —	Caernarfon/Gwynedd	171,808	99,017	555,452	166,085		208,504
1985 —	Gwent	175,953	109,213	598,647	211,710		491,369
1986 —	Anglesey/Gwynedd	176,117	104,258	671,380	162,350		216,996
1987 —	Carmarthen/Dyfed	186,527	124,948	696,711	175,684		386,315
1988 —	Radnor/Powys	193,998	118,559	747,779	143,248		295,693
1989 —	South Glamorgan	200,409	129,658	892,250	245,956		386,240
1990 —	Merioneth/Gwynedd	202,257	162,295	1,020,485	206,286		440,807
1991 —	Pembroke/Dyfed	219,053	157,449	1,034,522	122,260		343,970
1992 —	Montgomery/Powys	206,278	154,696	1,074,371	90,851		238,769
1993 —	West Glamorgan	218,915	169,946	1,098,056	165,443		232,523
1994 —	Clwyd	229,712	185,436	1,180,464	252,329		530,741
1995 —	Ceredigion/Dyfed	232,814	167,440	1,301,251	239,241		454,337
1996 —	Caernarfon	230,630	261,504	1,291,451	345,960		423,589
1997 —	Brecknock	226,413	257,664	1,312,825	310,041		258,419
1998 —	Anglesey	211,921	266,553	1,360,573	169,387		206,469
1999 —	Glamorgan	208,952	280,617	1,366,694	193,651		2,239,823
2000 —	Radnor	220,534	276,817	1,539,662	233,529		373,243
2001 —	[show cancelled]	–	277,161	–	–	390,000	11,724
2002 —	Old Monmouthshire	214,798	284,641	1,614,540	254,155		144,293

The surplus/deficit is before taking account of feature county contributions and life memberships.

RWAS OVERALL PROFIT AND LOSS: 1963–2000

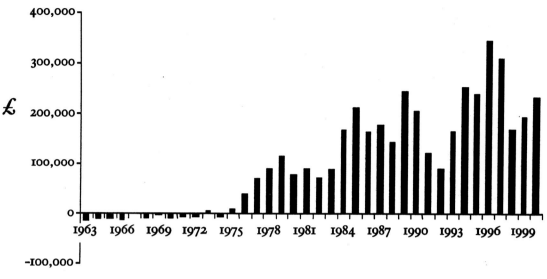

FIGURE I

SUMMER SHOW ATTENDANCES: 1963–2000

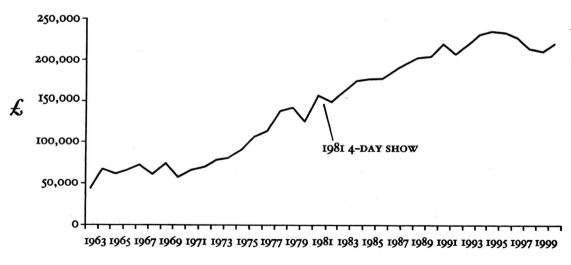

FIGURE 2

Comparative statement of entries 1963–2003

YEAR	TRADE STANDS	HORSE ENTRIES	CATTLE ENTRIES	SHEEP ENTRIES	PIG ENTRIES	GOAT ENTRIES
1963 – 3 days	293	643	434	443	125	
1964 – 3 days	292	748	441	490	120	
1965 – 3 days	308	755	424	553	139	
1966 – 3 days	298	778	420	486	114	
1967 – 3 days	296	904	506	516	98	
1968 – 3 days	284	1082	462	494	119	
1969 – 3 days	256	1105	368	525	101	
1970 – 3 days	264	1043	433	529	155	
1971 – 3 days	278	1047	455	538	145	
1972 – 3 days	325	1234	527	584	130	
1973 – 3 days	351	1125	574	590	–	
1974 – 3 days	389	1281	605	635	–	
1975 – 3 days	401	1183	561	614	–	
1976 – 3 days	465	1317	713	663	73	
1977 – 3 days	475	1382	603	696	62	154
1978 – 3 days	566	1481	636	711	71	192
1979 – 3 days	601	1633	661	683	69	219
1980 – 3 days	637	2249	580	858	68	244
1981 – 4 days	695	2107	574	785	38	188
1982 – 4 days	639	2031	554	873	67	259
1983 – 4 days	687	2098	630	845	48	255
1984 – 4 days	710	2429	636	1107	60	281
1985 – 4 days	765	2342	698	1157	42	279
1986 – 4 days	850	2388	619	1166	47	308
1987 – 4 days	860	2296	731	1244	41	198
1988 – 4 days	1000	2583	782	1450	32	236
1989 – 4 days	★	2960	897	1756	30	209
1990 – 4 days		3238	796	1830	40	227
1991 – 4 days		3298	798	1855	46	255
1992 – 4 days		3025	772	1923	47	241
1993 – 4 days		3243	808	2189	54	276
1994 – 4 days		3274	746	2096	71	209
1995 – 4 days		3275	821	2256	87	232
1996 – 4 days		3541	888	2270	80	211
1997 – 4 days		3667	820	2363	96	212
1998 – 4 days		2958	731	2417	77	227
1999 – 4 days		3279	824	2426	81	233
2000 – 4 days		3241	757	2480	81	223
2001 – 4 days		–	–	–	–	–
2002 – 4 days		3179	635	1554	44	127
2003 – 4 days		3630	816	2256	75	201

★The showground has been filled to its capacity of 1,000 stands from 1989 onwards.

Show venues, dates and presidents

YEAR	VENUE	DATE	PRESIDENT
1904	Aberystwyth	August	The Rt. Hon. The Earl of Powys
1905	Aberystwyth	August	The Rt. Hon. Lord Tredegar
1906	Aberystwyth	August	Sir Powlett Millbank
1907	Aberystwyth	August	The Rt. Hon. The Earl of Plymouth, PC, CB
1908	Aberystwyth	August	The Rt. Hon. Lord Harlech, KC, CB, TD
1909	Aberystwyth	August	The Rt. Hon. Sir. H. Aubrey Fletcher, PC, MP
1910	Llanelli	August	Sir John T. Dillwyn Llewelyn
1911	Welshpool	August	The Rt. Hon. The Earl of Powis
1912	Swansea	August	David Davies, Esq.
1913	Porthmadog	August	Sir Charles G. Assheton Smith
1914	Newport	August	The Rt. Hon. Lord Tredegar
1915–21	*no shows during First World War*		
1922	Wrexham	August	HRH The Prince of Wales, KG
1923	Welshpool	August	HRH The Prince of Wales, KG
1924	Bridgend	August	David Davies, MP
1925	Carmarthen	August	The Rt. Hon. Lord Kylsant, GCMG
1926	Bangor	August	The Rt. Hon. Lord Penrhyn
1927	Swansea	August	The Rt. Hon. The Earl of Dunraven, CB, DSO
1928	Wrexham	July	J. C. Read, JP
1929	Cardiff	July	The Rt. Hon. The Earl of Plymouth, PC, DL
1930	Caernarfon	July	The Rt. Hon. Lord Penrhyn
1931	Llanelli	July	Daniel Daniel
1932	Llandrindod Wells	July	Col. Sir Charles Venables-Llewelyn
1933	Aberystwyth	July	The Rt. Hon. The Earl of Lisburne
1934	Llandudno	July	The Rt. Hon. Lord Mostyn
1935	Haverfordwest	July	Sir Evan D. Jones, LLD
1936	Abergele	July	Col. H. C. L. Howard, CB, CMG, DSO
1937	Monmouth	July	Sir John C. E. Shelley-Rolls, DL
1938	Cardiff (Joint with RASE)	July	Reuben Haigh
1939	Caernarfon	July	The Rt. Hon. Lord Penrhyn
1940–46	*no shows during Second World War*		The Rt. Hon. Lord Penrhyn
1947	Carmarthen	August	HRH The Princess Elizabeth
1948	*no show – petrol rationing*		
1949	Swansea	August	Sir William A. Jenkins
1950	Abergele	July	Lt. Col. Sir Watkin Williams-Wynn, MFH
1951	Llandrindod Wells (Llanelwedd Site)	July	Brig. Sir Michael D. Venables-Llewelyn, MVO
1952	Caernarfon	July	Sir Michael Duff, K.St.J.
1953	Cardiff	July	Major C. G. Traherne, TD
1954	Machynlleth	July	Sir C. Bryner Jones, CB, CBE
1955	Haverfordwest	July	Col. Sir Thomas Meyrick
1956	Rhyl	July	Brig. H. S. K. Mainwaring
1957	Aberystwyth	July	Capt. J. Hext Lewes, OBE, RN(retd)
1958	Bangor	July	Sir Michael Duff, K.St.J.
1959	Port Talbot	July	Sir David M. Evans-Bevan
1960	Welshpool	July	The Rt. Hon. The Earl of Powis

1961	Llandeilo (Gelli-aur)	July	Col. Sir Grismond Philipps, CVO
1962	Wrexham	July	Col. J. C. Wynne Finch, CBE, MC
1963 ★	Llanelwedd *Radnor year*	July	Brig. Sir Michael D. Venables-Llewelyn, MVO
1964	Llanelwedd *Brecon year*	July	Bevington R. Gibbins, Esq.
1965	Llanelwedd *Anglesey year*	July	Dr J. T. Owen
1966	Llanelwedd *Pembroke year*	July	J. E. Gibby, OBE, FRAgS
1967	Llanelwedd *Cardiganshire year*	July	Dr Jenkin Alban Davies
1968	Llanelwedd *Merioneth year*	July	Col. John F. Williams-Wynne, CBE, DSO, FRAgS
1969	Llanelwedd *Glamorgan year*	July	A. B. Turnbull, OBE, FRAgS
1970	Llanelwedd *Caernarfon year*	July	Sir Charles Michael Robert Vivian Duff
1971	Llanelwedd *Monmouth year*	July	Col. Roderick Hill, DSO, K.St.J.
1972	Llanelwedd *Denbigh year*	July	Lt. Col. G. E. FitzHugh, OBE, TD, FRAgS
1973	Llanelwedd *Carmarthen year*	July	The late Mr Arwyn S. Lewis (Acting – Mrs Helen Lewis)
1974	Llanelwedd *Montgomery-Powys year*	July	T. Merfyn Jones, CBE
1975	Llanelwedd *South Glamorgan year*	July	Sir Julian Hodge
1976	Llanelwedd *Radnor-Powys year*	July	The Rt. Hon. Lord Gibson-Watt, MC, FRAgS
1977	Llanelwedd *West Glamorgan year*	July	The Rt. Hon. Lord Heycock
1978	Llanelwedd *Merioneth-Gwynedd year*	July	Meuric Rees, CBE, FRAgS
1979	Llanelwedd *Pembroke-Dyfed year*	July	The Lady Marion Philipps, FRAgS
1980	Llanelwedd *Brecon-Powys year*	July	Mrs R. W. P. Parry
1981	Llanelwedd *Denbigh-Clwyd year*	July	R. Gwynn Hughes
1982	Llanelwedd *Mid Glamorgan year*	July	T. M. Richards, MBE
1983	Llanelwedd *Cardigan-Dyfed year*	July	Geraint W. Howells, MP, FRAgS
1984	Llanelwedd *Caernarfon-Gwynedd year*	July	R. Pritchard Jones
1985	Llanelwedd *Monmouth-Gwent year*	July	Sir Harry Llewelyn, CBE
1986	Llanelwedd *Anglesey-Gwynedd year*	July	Tom Edwards, MBE
1987	Llanelwedd *Carmarthen-Dyfed year*	July	W. J. Hinds, MBE, FRAgS
1988	Llanelwedd *Radnor-Powys year*	July	V. W. Pugh, MBE, FRAgS
1989	Llanelwedd *South Glamorgan year*	July	Idwal Symonds
1990	Llanelwedd *Merioneth-Gwynedd year*	July	John E. Tudor, CBE, DL
1991	Llanelwedd *Pembroke-Dyfed year*	July	Peter J. Perkins, FRAgS
1992	Llanelwedd *Montgomery-Powys year*	July	The Hon. E. E. Islwyn Davies, CBE, FRAgS, DL
1993	Llanelwedd *West Glamorgan year*	July	Dr Gwyn Jones
1994	Llanelwedd *Clwyd year*	July	Michael Griffith
1995	Llanelwedd *Ceredigion year*	July	Tom Evans, MBE, FRAgS
1996	Llanelwedd *Caernarfon year*	July	D. L. Carey Evans, OBE, DL
1997	Llanelwedd *Brecknock year*	July	The Hon. Mrs Shân Legge-Bourke, LVO
1998	Llanelwedd *Anglesey year*	July	O. G. Thomas, DL, FRAgS
1999	Llanelwedd *Glamorgan year*	July	D. Hugh Thomas, CBE, K.St.J, DL
2000	Llanelwedd *Radnor year*	July	Robin Gibson-Watt, DL
2001	Llanelwedd *Monmouth year*	July	G. Stanley Thomas, OBE
2002	Llanelwedd *Monmouth year*	July	G. Stanley Thomas, OBE
2003	Llanelwedd *Merioneth year*	July	Robin Price, DL, ARAgS

★ Permanent site

FINANCE OR FINANCE AND GENERAL PURPOSES OR FINANCE AND EXECUTIVE COMMITTEE

1904–1907	Sir Powlett Milbank, Bt.	1975–1980	Edward Owen
1908–1944	David Davies (from 1932, Lord Davies, Llandinam)	1980–1986	Peter Perkins, FRAgS
		1987–1991	Lloyd FitzHugh, OBE, DL
1946–1948	Col. G. R. D. Harrison	1991–2000	H. G. Fetherstonhaugh, OBE
1949–1969	Lt. Col. G. E. FitzHugh, OBE, TD	2000 to date	John Vaughan, DL, FRAgS
1970–1975	A. M. Jones		

SCHEDULE (OR STOCK PRIZES) AND JUDGES SELECTION/LIVESTOCK COMMITTEE

1904–1908	Sir R. D. Green Price, Bt.	1968–1972	C. Austin Jenkins
1909–1914	Edward Green	1972–1976	Meuric Rees, CBE, FRAgS
1922–1944	Major David Davies (from 1932, Lord Davies, Llandinam)	1976–1998	Emlyn Kinsey Pugh, MBE, FRAgS
1946–1948	D. D. Williams	1998–2000	H. G. Fetherstonhaugh, OBE
1949–1968	Prof. J. E. Nichols	2000 to date	T. L. J. Clarke

SHOWYARDS WORKS COMMITTEE

1904–1909	D. Lloyd Lewis and John Roberts	1910–1914	David John and five others

FORESTRY COMMITTEE

1928–1936	Thomas Thomson	1969–1973	The Hon. Trevor O. Lewis
1937–1939	H. A. Hyde	1974–1982	M. L. Bourdillon
1946–1948	E. R. Puleston Jones	1982–1990	Col. J. D. Stephenson, MBE
1949–1955	Major J. D. D. Evans	1990 to date	Paul Raymond-Barker, FRICS
1956–1968	A. Lloyd O. Owen, CBE		

AMATEUR STOCK JUDGING AND DAIRYING COMPETITION

1931	G. Llewellin, Jnr.	1936	E. Hatfield (as above)
1932	E. Hatfield (Min. of Agric.)	1937	G. H. Purvis, FCS
1933	J. D. Davidson, ARCSI	1939	Isaac Jones
1934–1935	D. J. Morgan		

MACHINERY/ AND TRADESTANDS COMMITTEE

1949–1967	Prof. J. E. Nichols	1981–1986	C. J. Beynon
1967–1968	Lt. Col. J. J. Davis, OBE, TD, DL	1986–1998	Andrew Jones, MBE, FRAgS
1969–1971	R. E. Evans	1998 to date	Peter B. Evans
1972–1980	Peter Perkins, FRAgS		

HONORARY DIRECTOR'S OR SHOW ADMINISTRATION COMMITTEE

1951–1966	Lt. Col. G. E. FitzHugh, OBE, TD	1989–1994	Verney Pugh, OBE, FRAgS
1966–1969	Alan B. Turnbull, OBE, FRAgS	1994 to date	H. G. Fetherstonhaugh, OBE
1969–1989	Major David Fetherstonhaugh		

AREA DEMONSTRATION/COUNTRY PURSUITS/SPORTS AND COUNTRYSIDE CARE COMMITTEE

1965–1966	Hywel Evans, CBE	1996 to date	Glyn Sneade
1966–1995	Edward Griffiths, OBE, FRAgS		

EDITORIAL AND PUBLICITY COMMITTEE

1955	Major J. D. Gibson-Watt, MC, DL, FRAgS	1969–1979	R. H. Bowering, FRAgS
1956–1963	The Hon. Islwyn E. E. Davies, CBE, DL, FRAgS	1980–1982	Llywelyn Phillips
		1985–1986	Lloyd FitzHugh, OBE, DL
1964–1968	J. Llefelys Davies, CBE	1987–2000	John Vaughan, DL, FRAgS
1968	Sylvan Howell	2000 to date	W. Haydn Jones, MBE, FRAgS

HORTICULTURE COMMITTEE

1951–1954	The Rt. Hon. Lord Kenyon	1972–1978	Mrs K. Parry (later Stevenson)
1955–1959	The Most Hon. The Marquess of Anglesey, FSA	1978–1981	Ian S. Treseder
1960–1971	Mrs B. M. Austin Jenkins	1981 to date	Dr F. M. Slater, FRAgS

EDUCATION COMMITTEE

1963	T. H. Jones, CBE	1967–1976	Prof. Martin Jones
1964–1967	Dr Richard Phillips		

PRODUCE AND HONEY COMMITTEE

1955	H. A. Thomas, FAI	1967–1969	David Thomas
1956–1966	Principal D. S. Edwards (Llysfasi and Caernarfon)		

ROYAL WELSH NATIONAL HONEY SHOW COMMITTEE

April 1970–November 1977: Various members of the committee elected to the chair	November 1977 to date: James Thomas, CBE, FRASC

PRODUCE AND HANDICRAFTS

1970–1975	David Thomas	1979 to date	James Thomas, CBE, FRASC
1976–1979	John Lewis		

CANINE COMMITTEE

1964–1981	William Prytherch	1982 to date	Trevor M. Evans

FUR AND FEATHER SECTION

1967–1976	David Thomas	1976 to date	James Thomas, CBE, FRASC

SHEEP SHEARING

1960–1961	W. J. Constable	1990–1991	J. T. Davies
1962–1967	J. Howard Bevan	1992–1993	T. A. Evans
1968–1974	T. E. Lewis	1994–1995	J. A. Davies
1975–1978	J. E. Roberts	1996–1997	B. Jones
1979–1981	E. L. Evans	1998–1999	M. R. David
1982–1986	Michael J. Evans, ARAgS	2000–2001	E. Evans
1986–1988	J. L. Davies	2002 to date	G. R. Jones
1988–1989	B. S. Williams		

MISS ROYAL WELSH

1970–1972	The Dowager Viscountess Chetwynd	1974–1988	Mrs N. S. K. Pugh
1972–1974	Miss Lorraine Jones	1988–1991	Miss Delyth Lewis
1974	MissVera Jones, MBE	1991 to date	Mrs Delyth (née Lewis) Jenkins

SPONSORSHIP

| 1980–1989 | Alan B. Turnbull, OBE, FRAgS | 1989 to date | W. L. S. Clay |

HORSE SHOEING/ORNAMENTAL IRONWORK

| 1976–1993 | William Jones | 1994 to date | Stephen K. Pugh |

MERCHANDISE

| 1992–1994 | Mrs Mari Edwards | 1998 to date | Mrs Sarah Froggatt |
| 1995–1998 | Mrs Barbara Morgan | | |

INVESTMENT

| 1987–1999 | The Hon. Islwyn Davies, CBE, DL, FRAgS | 2000 to date | W. Emrys Evans, CBE, FCIB |

WINTER FAIR

| 1987–1996 | D. Hanks, MBE | 1997 to date | H. G. Hughes |

PROGRAMME

| 1970–1990 | Alan B. Turnbull, OBE, FRAgS | 1991 to date | D. R. Thomas |

MEMBERSHIP

| 1977–1987 | Tudor Davies | 2000 to date | John Rees |
| 1988–1999 | Dewi M. Thomas, MBE, ARAgS | | |

PLANNING/DEVELOPMENT

| 1985–1988 | A. J. B. Ratcliffe, OBE | 1998 to date | W. Emrys Evans, CBE, FCIB |
| 1989–1997 | Lloyd FitzHugh, OBE, DL | | |

JUNIOR COMMITTEE

| 1988 to closure | James Thomas, CBE, FRASC |

Chief Officials

SECRETARY MANAGER AND HONORARY DIRECTOR
1904–1908   Lewes T. Loveden Pryse

SECRETARY TO COUNCIL
1908–February 1909   Prof. C. Bryner Jones, CBE
March 1909–February 1912 Robert Roberts
February 1912–1914   Thomas Whitfield, Jnr.

SHOW SECRETARY
1904–1914   Appointed representative from staff of Messrs. Thomas Whitfield
             and Co., Estate Agents, Shrewsbury.
             Nominated representatives: Walter Williams, Robert Roberts and
             Thomas Whitfield, Jnr.

ASSISTANT SECRETARY
1922–1948   Walter Williams

GENERAL SECRETARY
1927–1948   Capt. T. A. Howson (known as 'Lancastrian' as a journalist)
1948–1973   John Arthur George (Secretary/Manager from 1966), MBE, FRAgS

CHIEF EXECUTIVE
January 1974–May 1975   Philip S. Phillips
June 1975–September 1977 (Secretary/Manager) John Wigley, OBE, FRAgS
September 1977–August 1984 (Chief Executive) John Wigley, OBE, FRAgS
September 1984 to date (Chief Executive)   David Walters, FRAgS

SECRETARY TO THE RWAS
April 1987 to date   Peter Guthrie

ESTATES OFFICER
1981 to date   Brian Waller

SPONSORSHIP SECRETARY
1980–1988   Lt. Col. Desmond Evans
1989–1994   Gordon Hamer

SENIOR VETERINARY OFFICER
1904–1909   R. D. Williams
1910–1914   Richard Jones
1947–1968   T. H. Jones
1968–1975   F. V. John
1975–1993   T. Boundy, MBE, FRAgS
1993 to date D. E. Bowen

HONORARY TREASURERS
1904–1930   Arthur Jones
1930–1946   R. H. Thomas
1946–1961   J. E. Rees
1962–1966   J. Smith Davies
1967   (Joint) J.Smith Davies and D. Gwynne Hughes
1968   D.Gwynne Hughes
1969   (Joint) D. Gwynne Hughes and Richard H. Moseley, FFA
1970 to date Richard H. Moseley, FFA

PUBLIC RELATIONS CONSULTANT
1971 to date   John Kendall, FRAgS

93. Members of the Board of Management, 1997–9: (see key below) 1. John V. Williams, 2. Emlyn Kinsey Pugh, 3. A. J. B. Radcliffe, 4. Edward C. O. Owen, 5. Leslie T. Jones, 6. Peter D. Guthrie (secretary), 7. John Vaughan, 8. Verney W. Pugh, 9. Robin Price, 10. R. H. Moseley, 11. H. George Hughes, 12. Peter J. Perkins, 13. James Thomas, 14. Leslie R. Williams, 15. Llewelyn Evans, 16. Tudor J. Davies, 17. Desmond R. Evans, 18. Harry Harries, 19. John Rees, 20. J. Robin Gill, 21. Peter Sturrock, 22. Dewi M. Thomas, 23. W. I. Cyril Davies, 24. Cynthia Higgon, 25. Fred M. Slater, 26. Susan Jones, 27. Meuric Rees, 28. W. Emrys Evans, 29. D. Hugh Thomas (1999 president), 30. Harry G. Fetherstonhaugh, 31. David Walters (chief executive). Missing from this photograph are: Islwyn E. Davies, Derick H. Hanks, Emrys L. Griffiths, John Kendall, Peter B. Evans, Rosemarie Harris.

From its earliest years the officials of the Society were conscious of the need to keep a faithful record of the Society's deliberations and activities. That admirable concern has certainly facilitated the writing of this centenary history. An added bonus for the historian – if a burden for the office staff – has been the many committees charged with running the Society, whose discussions and resolutions at their regular meetings were recorded. Accordingly, much of the text is based on the minutes of the Society housed at the Department of Manuscripts at the National Library of Wales, Aberystwyth, and, those from the late 1950s, at the Society's archive at Llanelwedd. I have also made use of information contained in the long run of the Society's *Journal*; the latter's withdrawal between 1910 and 1922, however, left an unfortunate dearth of information that had to be filled by both the official minutes and the contemporary newspapers. Press reports were invaluable for shedding light on every phase of the Society's history. Of particular help in unravelling the early difficult relationships between the fledgling Society and the existing local agricultural societies was the Aberystwyth-based *Cambrian News*. The other newspaper deeply quarried was the *Western Mail*, which from the beginning carried full and sympathetic reports on the Society's activities. Apt citations from *Y Cymro* are also included. Information about the founding of the Society was also contained in the Gogerddan estate archive (box 66) housed at the National Library of Wales, wherein the correspondence to and from Lewes T. Loveden Pryse proved especially helpful. Much valuable information was likewise furnished from the personal papers of Arthur George, the official at the helm of the Society between 1948 and 1973. Dr Richard George kindly placed his father's papers at my disposal. George's successor between the years 1975 and 1984 was John Wigley, who in an unpublished manuscript housed at Llanelwedd drew together an amazing amount of information about the Society's history down to the late 1980s. Without access to this rich document my task of writing this centenary history would have been immeasurably more difficult.

I have also benefited from interviews granted me by persons knowledgeable about the Society's affairs, including the Honourable Islwyn Davies, Lord David Gibson-Watt, Tudor Davies, Graham Rees, David Lloyd and Dr W. Emrys Evans. In response to appeals made on radio a number of persons kindly provided information and personal reminiscences of their visits to the show. In so far as the photographs are concerned, besides the many drawn from the Society's archive at Llanelwedd and from its journals some were obtained from the Geoff Charles collection at the Department of Maps and Prints at the National Library of Wales.

While the decision was made at the outset not to employ the normal scaffolding of footnotes and a lengthy bibliography I would nevertheless want to draw the reader's attention to certain books and articles which have furnished the necessary historical context:

Ashby, A. W., 'The Agricultural Depression in Wales', *The Welsh Outlook*, 16 (1929).

Ashby, A. W., 'Some Characteristics of Welsh Farming', *The Welsh Outlook*, 20 (1933).

Ashby, A. W., 'The Peasant Agriculture of Wales', *Welsh Review*, 3 (1944).

Ashby, A. W. and Evans, I. L., *The Agriculture of Wales and Monmouthshire* (Cardiff: University of Wales Press, 1944).

Davies, John, 'The End of the Great Estates and the Rise of Freehold Farming in Wales', *Welsh History Review*, 7 (1974–5).

Davies, Walter, *The Agriculture and Domestic Economy of South Wales* (2 vols, London, 1815).

Davies, Wynne, *The Welsh Cob* (J. A. Allen, 1998).

Davies, Wynne, *One Hundred Glorious Years: The Welsh Pony and Cob Society, 1901–2001* (The Welsh Pony and Cob Society, 2001).

Edmunds, H., 'History of the Brecknockshire Agricultural Society, 1755–1955', *Brycheiniog*, 2–3 (1956–7).

Goddard, Nicholas, *Harvests of Change: The Royal Agricultural Society of England, 1838–1988* (London: Quiller Press, 1988).

Howell, D. W., *Land and People in Nineteenth-century Wales* (London: Routledge & Kegan Paul, 1977).

Howell, D. W., 'Farming in Pembrokeshire, 1815–1974', in D. W. Howell (ed.), *Modern Pembrokeshire, 1815–1974*, vol. 4, *Pembrokeshire County History* (Haverfordwest: Pembrokeshire Historical Society, 1993).

Hudson, Kenneth, *The Bath and West: A Bicentenary History* (Bradford-on-Avon: Moonraker Press, 1966).

Lewis, John, *Three into One: The Three Counties Agricultural Society, 1797–1997* (Baron Press, 1996).

McCreary, Alf, *On with the Show: 100 Years at Balmoral* (Royal Ulster Agricultural Society, 1996).

Martin, John, *The Development of Modern Agriculture: British Farming since 1931* (Basingstoke: Macmillan, 2000).

Moore-Colyer, R. J., 'The Pryce Family of Gogerddan and the Decline of a Great Estate, 1800–1960', *Welsh History Review*, 9 (1978–9).

Moore-Colyer, R. J., 'Early Agricultural Societies in South Wales', *Welsh History Review*, 15 (1986).

Moore-Colyer, R. J., 'Farming in Depression: Wales between the Wars, 1919–39', *Agricultural History Review*, 46, part 2 (1998).

Rees, Derek, *Rings and Rosettes: The History of the Pembrokeshire Agricultural Society 1784–1977* (printed by Gomer Press, Llandysul, 1977).

Whetham, Edith, *The Agrarian History of England and Wales*, viii *1914–1939* (Cambridge: Cambridge University Press, 1978).

Williams, L. J., *Digest of Welsh Historical Statistics*, 2 vols (Cardiff: Welsh Office, 1985).